Sharks
The Perfect Predators

SHARKS
THE PERFECT PREDATORS

Alessandro De Maddalena

Photographs by Vittorio Gabriotti
and Walter Heim

JACANA

First published by Jacana Media (Pty) Ltd in 2008

10 Orange Street
Sunnyside
Auckland Park 2092
South Africa
+2711 628 3200
www.jacana.co.za

ISBN 978-1-77009-559-5

Front cover photograph:
   This shortfin mako (*Isurus oxyrinchus*) was found with copepods (parasitic
   crustaceans) attached to the mouth and first dorsal fin (photograph by Walter Heim)
Back cover photograph:
   A blue shark (*Prionace glauca*) swimming at the sea surface off San Diego,
   California (photograph by Walter Heim)

Set in Sabon 10/14pt
Printed by Imago Productions F.E. Pte Ltd.
Job No. 000614

See a complete list of Jacana titles at www.jacana.co.za

This book is dedicated to
my wife Alessandra and
my son Antonio.

Contents

A hawksbill turtle (*Eretmochelys imbricata*) swims in the Maldive waters. The varied diet of some large sharks includes marine reptiles, such as sea turtles (photograph by Vittorio Gabriotti).

FOREWORD

Alessandro De Maddalena is driven by his passion for sharks and a quest to learn more about them. He has written numerous scientific publications and several books on sharks. It is a great pleasure for me to write the Foreword to his latest book, in which he so aptly describes the shark as being the perfect predator.

The wide variety of prey eaten by many shark species is well documented. It is common knowledge, particularly in the case of the tiger sharks, that the diet does occasionally include terrestrial animals and a stunning variety of inedible or indigestible items. Undeniably more intriguing, but not as well known, is the manner in which the shark tracks down its prey. The author provides a stimulating insight into the factors that may motivate a shark in its quest for food, and the tactics it utilises to hunt down its prey.

The author concedes that there are still large gaps in our knowledge of shark predatory strategies. This is because most species are very difficult to observe, let alone study in the wild, and because many of their activities are performed at night. I am sure that this publication will stimulate further academic discussion and research in this field. More importantly, it will arm the enormous number of amateur shark watchers, who travel the globe to dive with sharks, with the necessary tools and insight to be able to make meaningful observations of sharks in their pursuit of prey. I would urge readers to submit their observations, using the reporting form provided in Appendix I.

The book is greatly enhanced by numerous photographs, captured by Vittorio Gabriotti and Walter Heim. I am confident that it will inspire its readers to a greater level of respect and admiration for a group of animals that were once largely feared and despised.

Geremy Cliff
Natal Sharks Board
Umhlanga Rocks
South Africa
November 2007

PREFACE

Sharks are especially attractive animals. No other marine creature is surrounded by as much mystery and fear. Moreover, the great size of some species, such as the whale shark and great white shark, has long fascinated people. Sharks are beautiful, graceful and powerful, but they are also the sea's most feared predators, and many people think of them as voracious and ferocious monsters, consuming anything they encounter: this is far from the truth. Most sharks are difficult to keep in captivity because under these conditions they often refuse to feed, and die. Even in their natural habitat most sharks do not feed every day. Many shark species are highly specialised predators with preferred food sources.

There are many gaps in our knowledge of sharks. Most studies concentrate on the few species considered to have commercial importance. Many species of sharks are difficult to study in the wild, being uncommon or elusive and timid. But today, our knowledge is much greater than it was only a few years ago. Facts have replaced the myths that were associated with these creatures.

Sharks need to capture, ingest and digest food as efficiently as possible: morphological, anatomical and behavioural adaptations enable them to do this. They often capture, kill and eat their prey in a spectacular way. Some of the largest sharks can also swallow inedible items. These fish have evolved a wide variety of predatory strategies, but no shark relies exclusively on a single tactic to capture its prey.

Some sharks are among the few remaining organisms capable of preying on our own species. However, despite their bad reputation, most sharks are harmless to humans, and even dangerous species rarely attack unless provoked. We know that most attacks on humans are not motivated by hunger. Most species are timid and inoffensive. However, all sharks should be treated with caution and respect.

Until recently we have known little about the behaviour of these efficient predators; we have only begun to explore the world of sharks. Slowly we are filling the gaps in our knowledge about their predatory tactics and diet. An aim of this book is to stimulate further studies on the subject.

This book is based on the studies conducted by numerous researchers (the number of scientific papers and articles on sharks is enormous). What do sharks eat? Which animals are especially vulnerable to shark predation? When do sharks feed? Where do sharks catch their prey? How do sharks find their prey? Are all sharks dangerous to humans? Why do some sharks attack humans? What

A shortfin mako shark (*Isurus oxyrinchus*) (photograph by Walter Heim)

predatory tactics do sharks use? What happens when two sharks try to feed on the same prey? This book provides answers to these and many other questions. It has been prepared primarily for marine life enthusiasts, then for biologists and zoologists. It is a scientific book, but I have tried to simplify the more complex parts and make them easily comprehensible to the general reader.

The text is scientific and detailed, but all technical terms are explained simply. In order to avoid repetition, I have cited the Latin name of a given species once per chapter and thereafter used its common name. I have avoided the use of parenthetical source citations, which in scientific papers enable the researcher to refer to the source of information; but wherever possible the source of specific data has been identified in the text. Readers interested in knowing more about sharks will find many useful references in the Bibliography at the end of the book.

The text is supported by many pictures, most of which have been prepared especially for this book. I have contributed the photographs of shark jaws and teeth; Vittorio Gabriotti and Walter Heim have provided wonderful photographs of live animals; a few additional rare images have been obtained from other sources.

Alessandro De Maddalena
Italian Ichthyological Society
Milan, Italy
January 2007

Acknowledgements

Many people and institutions have contributed to the data, information and pictures in this book.

I must pay special homage to Walter and Beverly Heim (San Diego, California, USA), who took the time to read and comment on the entire manuscript.

Very special thanks to Geremy Cliff (Natal Sharks Board, Umhlanga Rocks, South Africa) for providing the Foreword.

Special thanks to the two photographers who contributed wonderful pictures to the publication: Vittorio Gabriotti and Walter Heim.

I thank the following people for freely sharing their observations and for their assistance in assembling material for this book:

Nicola Allegri (Italy)

Sean S Anderson (Department of Organismal Biology, Ecology and Evolution, University of California, Los Angeles, USA)

Harald Baensch (SharkProject, Munich, Germany)

Joan Barrull (Laboratorio Vertebrats, Secciò Ictiologia, Museu de Zoologia, Barcelona, Spain)

Michèle Bruni (Musée Océanographique de Monaco, Monaco, Principauté de Monaco)

David Catania (Department of Ichthyology, California Academy of Sciences, San Francisco, California, USA)

Antonio Celona (Aquastudio Research Institute, Messina, Italy)

Tony Chess (Piercy, California, USA)

Geremy Cliff (Natal Sharks Board, Umhlanga Rocks, South Africa)

Ralph Collier (Shark Research Committee, Van Nuys, California, USA)

Giorgia Comparetto (Necton Marine Research Society, Catania, Italy)

Emiliano D'Andrea (Necton Marine Research Society, Catania, Italy)

Shawn Dick (Aquatic Release Conservation, ARC Dehooker, Ormond Beach, Florida, USA)

David A Ebert (Pacific Shark Research Center, Moss Landing Marine Laboratories, Moss Landing, USA)

Andrew Fox (Rodney Fox Great White Shark Expeditions, Adelaide, Australia)

Rodney Fox (Rodney Fox Great White Shark Expeditions, Adelaide, Australia)

Roberta Gabriotti (Tritone Sub, Brescia, Italy)

Vittorio Gabriotti (Tritone Sub, Brescia, Italy)

Chris Gotschalk (Marine Science Institute, University of California Santa Barbara, Santa Barbara, California, USA)

Eric G Haenni (University of Saint Francis, Fort Wayne, Indiana, USA)

Jessica Heim (San Diego, California, USA)

Richard Herrmann (Poway, California, USA)

Rohan Holt (Countryside Council for Wales, Gwynedd, Wales, United Kingdom)
Mick Jansen
Stephen M Kajiura (Ecology and Evolutionary Biology, University of California, Irvine, USA)
Tadashi Kubota (Department of Fisheries, Tokai University, Japan)
Wolfgang Leander (www.oceanicdreams.com, Cochabamba, Bolivia)
Christopher G Lowe (Department of Biological Science, California State University, Long Beach, USA)
Renato Malandra (Mercato Ittico, Milano, Italy)
Richard Aidan Martin (ReefQuest Centre for Shark Research, Vancouver, Canada)
Isabel Mate (Laboratorio Vertebrats, Secciò Ictiologia, Museu de Zoologia, Barcelona, Spain)
Jirí Mlíkovský (Department of Zoology, National Museum, Prague, Czech Republic)
Marcela Moisset (Cordoba, Argentina)
Juan Antonio Moreno (Facultad de Ciencias Biológicas de la Universidad Complutense de Madrid, Villacastín, Spain)
Christopher Parsons (Tennessee Aquarium, Chattanooga, Tennessee, USA)
Ian 'Pato' Paterson (Rodney Fox Great White Shark Expeditions, Adelaide, Australia)
Claudio Perotti (Brescia, Italy)
Luigi Piscitelli (Società Ittiologica Italiana, Milano, Italy)
Rachel Powell (Rodney Fox Great White Shark Expeditions, Adelaide, Australia)
Antonella Preti (Southwest Fisheries Science Center, NMFS, San Diego, California, USA)
Radek Šanda (Department of Zoology, National Museum, Prague, Czech Republic)
Jeff Shindle (California, USA)
Susan E Smith (National Marine Fisheries Service, La Jolla, California, USA)
Mommo Solina (the tuna trap in Bonagìa, Bonagìa, Italy)
Pushpa Soobraya (Natal Sharks Board, Umhlanga Rocks, South Africa)
Southeast Fisheries Science Center (NOAA Fisheries Service, Miami, Florida, USA)
John D Stevens (CSIRO Division of Marine Research, Hobart, Tasmania, Australia)
The Sydney Aquarium
Joost Wenderich (Keep Smiling Diving Divecenter, Reeuwijk, The Netherlands)
WildAid (San Francisco, USA)
Alberto Zanoli (Milan, Italy)
Phil Zerofski (SEACAMP San Diego, San Diego, California, USA)
Marco Zuffa (Museo Archeologico 'Luigi Donini', Ozzano dell'Emilia, Italy)

For their help, support and friendship, my sincere gratitude goes to Alessandra Baldi, Antonio De Maddalena, Roberta Gabriotti, the Italian Ichthyological Society, and the Mediterranean Shark Research Group. I also thank Sauro Baldi for providing information on the Maya origin of the word 'shark'.

My gratitude also goes to the publisher for their assistance.

THE AUTHOR

Alessandro De Maddalena (Milan, 1970) is one of the major shark experts in Europe. He is the curator of the Italian Great White Shark Data Bank, the President of the Italian Ichthyological Society, and a founding member of the Mediterranean Shark Research Group.

His research findings have been published in numerous scientific publications including *Annales Series historia naturalis*, *Museologia Scientifica*, *Annali del Museo Civico di Storia Naturale di Genova*, *Bollettino del Museo civico di Storia naturale di Venezia*, *Thalassia Salentina*, *Biljeske – Notes*, *Journal of the National Museum of Prague*, *South African Journal of Science*, *Marine Life*, and the *Latin American Journal of Aquatic Mammals*.

He is also the author of nine books on sharks: *Squali delle Acque Italiane. Guida Sintetica al Riconoscimento* (Ireco, 2001); *Lo Squalo Bianco nei Mari d'Italia* (Ireco, 2002); *Sharks of the Adriatic Sea* (Knjiznica Annales Majora, 2004, co-authored with Lovrenc Lipej and Alen Soldo); *Mako Sharks* (Krieger Publishing, 2005, co-authored with Antonella Preti and Robert Smith); *Haie im Mittelmeer* (Kosmos Verlag, 2005, co-authored with Harald Baensch); *Guida all'identificazione in mare dei grandi animali del Mediterraneo* (Rivista Marittima, 2005, co-authored with Antonio Celona); *Great White Sharks Preserved in European Museums* (Cambridge Scholars Publishing, 2007); *Sharks of the Pacific Northwest (including Oregon, Washington, British Columbia and Alaska)* (Harbour Publishing, 2007, co-authored with Antonella Preti and Tarik Polansky); and *25 Haie* (Verlagsedition nullzeit.eu, 2007).

Considered one of the world's best nature illustrators, Alessandro De Maddalena has produced over seven hundred illustrations of sharks and cetaceans. Among the books that feature his illustrations are *Les requins des cotes françaises* by Bernard Séret (Editions Ouest-France, 1999), *Field Guide to the Great White Shark* by R Aidan Martin (ReefQuest Centre for Shark Research, 2003) and *Shark Attacks of the Twentieth Century from the Pacific Coast of North America* by Ralph S Collier (Scientia Publishing, 2003).

His articles and illustrations have appeared in many wildlife magazines, including *The World and I*, *Dive New Zealand*, *Dive Pacific*, *Annales*, *Biologie in unserer Zeit*, *Unterwasser*, *Tauchen*, *Duiken*, *Apnéa*, *Plongeurs International*, *Océanorama*, *Enviromagazin*, *Mondo Sommerso*, *Il Pesce*, *EuroFishmarket*, *Aqva*, *Quark*, *Airone* and *Rivista Marittima*. He has had illustrations on exhibit at the Oceanographic Museum in Monaco, the World Festival of Underwater Pictures in Antibes – Juan-Les-Pins, Nausicaä – the

French National Sea Experience Centre in Boulogne-sur-Mer, the Aquarium of Milan, the Whale Shark Museum in Philippines, and the Museum of Marine Biology in Porto Cesareo.

Contact details:
Dr Alessandro De Maddalena
Italian Ichthyological Society
via L Ariosto 4, 20145 Milano, Italy
Tel. 39/0248021454
E-mail: a-demaddalena@tiscali.it
Web site: www.geocities.com/demaddalena_a/demaddalena.html

Alessandro De Maddalena displaying the jaws of an (approximately) three-metre oceanic whitetip shark (*Carcharhinus longimanus*) (photograph by Alessandro De Maddalena).

The Photographers

Vittorio Gabriotti (photograph by Roberta Gabriotti)

Vittorio Gabriotti

Vittorio Gabriotti (Brescia, 1964) is an architect and a professional marine and wildlife photographer. He has produced photographs of sharks (including great white sharks in South Australian waters) and numerous other marine animals in his travels around the world and on visits to the Mediterranean Sea, the Indian Ocean, the Red Sea and the Pacific Ocean.

His pictures have appeared in numerous Italian magazines, including *Aqva*, *Mondo Sommerso*, *Il Subacqueo*, *Sub*, *Il Pesce*, *Pescasub*, *Bollettino del Museo civico di Storia Naturale di Venezia* and *Dive New Zealand*. He is co-author of the book *In fondo al lago. Guida alle immersioni nel Lago di Garda* (Provincia di Brescia, 2004), a photographic guide to diving in the Lake of Garda. He has been diving since 1978 and is also a scuba diving instructor. He is a founding member of the Italian Ichthyological Society, and a member of the Mediterranean Shark Research Group.

Contact details:
Vittorio Gabriotti
Via Filippo Corridoni 5, 25128 Brescia, Italy
E-mail: info@grandesqualobianco.com
Website: www.grandesqualobianco.com

WALTER HEIM

Walter Heim (San Diego, 1957) is a mechanical
engineer and amateur photographer. He has been
fishing and diving the water off San Diego for over
thirty years. Walter became a volunteer shark
tagger in the late 1990s, and has tagged over two
hundred mako sharks. He started photographing
sharks in 2001, and with his daughter Jessica
has developed techniques to photograph mako
and blue sharks at close range. His pictures have
appeared in many wildlife magazines, including
Shark Diver Magazine, *Dive New Zealand*,
Airone and *Océans*. Amongst the books that
feature his photographs are *Mako Sharks* by
Alessandro De Maddalena, Antonella Preti and
Robert Smith (Krieger Publishing, 2005) and *Haie
im Mittelmeer* by Alessandro De Maddalena and
Harald Baensch (Kosmos Verlag, 2005).

Walter Heim (photograph by
Jessica Heim)

Contact details:
Walter Heim
San Diego, California, USA
E-mail: wheim@sdccu.net

THE SHARK: PORTRAIT OF A PERFECT PREDATOR

The origin of the word 'shark' could be 'Xok', a word in the Maya language. According to american Mayanist Tom Jones, the 'Xok' is specifically the bull shark (*Carcharhinus leucas*), a very dangerous and well-known species that also frequents freshwater, including Latin American rivers. English seamen learned this word in the 17th century, during an expedition to the Bay of Campeche, Mexico. When they returned to England, the word was added to the dictionary as 'shark'.

The great white shark (*Carcharodon carcharias*), the perfect predator (photograph by Vittorio Gabriotti)

CLASSIFICATION

Sharks belong to the phylum Chordata, the subphylum Vertebrata, the class Chondrichthyes, the subclass Elasmobranchii, the superorder Selachimorpha. Sharks, rays and chimaeras are Chondrichthyes or cartilaginous fishes.

These animals have skeletons composed of cartilage (the only bony tissues are found in their teeth and scales), while Osteichthyes or bony fishes (also called teleosts) have skeletons made of bone. Cartilage is a light and flexible connective tissue that also forms the skeleton of the human embryo, but is predominantly replaced by bone during development (cartilage persists in many areas of the adult human being, such as the nose, the ears, and the articular surfaces of joints).

Sharks are classified into eight orders: Hexanchiformes (frilled and cow sharks), Squaliformes (dogfish sharks), Pristiophoriformes (saw sharks), Squatiniformes (angelsharks), Heterodontiformes (bullhead sharks), Orectolobiformes (carpet sharks), Lamniformes (mackerel sharks) and Carcharhiniformes (ground sharks). These orders are divided into 34 families that include 479 species of sharks (this number includes a few dubious species that may not be valid). While this is the present situation, the classification of sharks is continually being revised. With new discoveries and an increasing knowledge of shark morphology, this classification undoubtedly will continue to change.

The bluespotted ribbontail ray (*Taeniura lymma*). Sharks and rays are cartilaginous fish and form the subclass Elasmobranchii (photograph by Vittorio Gabriotti).

EVOLUTION

Sharks are a very ancient group. They arose some 400 million years ago, between the Silurian and the Early Devonian periods. They almost certainly evolved from the placoderms, a group of extinct armoured bony fish (the earliest branch of the jawed fish).

Although shark vertebrae and tiny dermal denticles (see 'Skin', page 16) can occasionally be preserved as fossils owing to their partial calcification, complete skeletons are preserved only in very rare cases. However, while the cartilaginous skeleton rapidly disintegrates after death, teeth fossilise easily and are usually the only remains of many extinct species.

Although somewhat fragmentary in nature, there is significant evidence of shark evolutionary history. Sharks have remained virtually unchanged for the past 100 million years. The retention of a similar body form by prehistoric and modern sharks suggests that early in their evolutionary history these fish developed morphological and anatomical characteristics that make them successful animals, well adapted to their environment. Consequently, sharks are highly evolved animals.

Otodus obliquus fossil tooth. This large shark reached lengths of 6 metres, and its teeth were up to 8cm long (photograph by Alessandro De Maddalena).

The huge basking shark (*Cetorhinus maximus*) may grow to 14m (photograph by Chris Gotschalk).

SIZE

There are great differences in size among shark species. Most sharks are small and do not reach 1.5m. There are 11 gigantic species, exceeding 4.5m in length: whale shark (*Rhincodon typus*), basking shark (*Cetorhinus maximus*), great white shark (*Carcharodon carcharias*), tiger shark (*Galeocerdo cuvier*), Greenland shark (*Somniosus microcephalus*), Pacific sleeper shark (*Somniosus pacificus*), common thresher shark (*Alopias vulpinus*), bigeye thresher (*Alopias superciliosus*), great hammerhead (*Sphyrna mokarran*), megamouth shark (*Megachasma pelagios*) and bluntnose sixgill shark (*Hexanchus griseus*). One of these, the megamouth shark, was discovered recently – in 1976. The huge whale shark may grow to 20m (length reported for a specimen caught in Taiwan waters) and is the world's largest fish.

Excluding the plankton-eaters, the largest predator shark is the great white shark: its maximum size is recorded at 6.6m, as evidenced by specimens caught in the Messina Strait, Italy, and Maltese waters, but Spanish shark specialist Juan Antonio Moreno estimated a total length of over 8m for a specimen caught off Dakar, Senegal. Female sharks attain larger sizes than males.

MORPHOLOGY AND SWIMMING

Sharks usually have a streamlined body, a long flattened snout, a ventral parabolic mouth, and an asymmetric caudal fin with the upper lobe noticeably longer than the lower. Body form varies considerably among species, and is related to the shark's way of life: for example, body form is long and slender in benthic catsharks (family Scyliorhinidae), large and conico-cylindrical in pelagic fast-swimming mackerel sharks (family Lamnidae), and strongly flattened in benthic angelsharks (family Squatinidae) and saw sharks (family Pristiophoridae). 'Pelagic' refers to all species living in the open sea, while 'benthic' defines species living most of the time on the sea bottom.

Most sharks have eight fins: two pectoral, two pelvic, first dorsal, second dorsal, anal and caudal fins. The caudal fin is used for propulsion: a shark swims by moving its caudal fin from side to side. The longer upper caudal fin lobe drives the shark down during swimming, but this is balanced by lift generated from

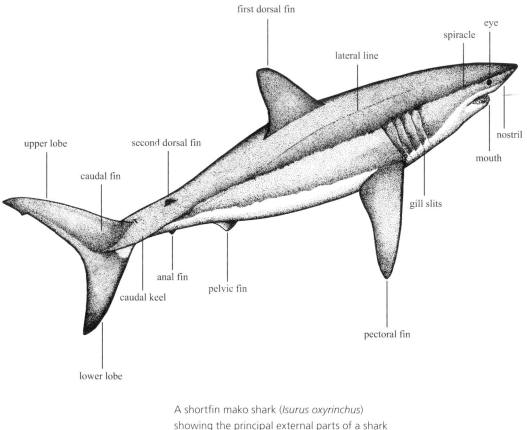

A shortfin mako shark (*Isurus oxyrinchus*)
showing the principal external parts of a shark
(drawing by Alessandro De Maddalena)

14

the flattened head and the almost horizontal wide pectoral fins. One function of the wide head of the hammerhead shark (family Sphyrnidae) is to provide more lift (this also enables these sharks to make fast vertical movements, giving them greater manouverability). In some benthic species, the lower lobe is almost absent, while in some fast-swimming species (mackerel sharks) the upper and lower lobes are nearly of equal size, and the snout is conical.

For improved hydrodynamic design, many sharks have lateral keels on the sides of the caudal peduncle (the caudal peduncle is slightly or strongly flattened and laterally expanded). The fastest species may be the blue shark (*Prionace glauca*) and the shortfin mako (*Isurus oxyrinchus*), with maximum speeds of 69 and 35-56kmh respectively; the average speed is much lower. Some species have a pronounced spine immediately anterior to each dorsal fin, used for defence.

RESPIRATION

Sharks employ gills for respiration. The mouth of the shark leads to the buccal cavity and to the pharynx where the gills are located. Oxygen is extracted from the sea water by highly vascularised membranes called gill lamellae, while

Sharks have five to seven pairs of gill slits. Bony fish gills are covered by a flap called an operculum; shark gill slits are uncovered (photograph by Walter Heim).

carbon dioxide is expelled to the sea water. Most sharks have five pairs of gill slits, except species of the small order Hexanchiformes (five species) and the genus *Pliotrema* (one species), which have six or seven gill slits. While bony fish gills are covered by a flap called an operculum, the gill slits of the shark are uncovered, and consequently they are external.

Moreover, many sharks have two small openings situated behind or below the eyes, one per side: these are the spiracles or rudimentary gill openings. The spiracles are used instead of the mouth as an entrance for water, especially in sharks that live on the sea bottom; therefore the spiracles are larger in benthic species and smaller or absent in pelagic species. Pelagic sharks, the most active species, require larger amounts of oxygen; therefore they have to swim constantly to stay alive.

BUOYANCY

Most bony fish have a gas bladder, a gas-filled sac located in the upper part of the body cavity used to offset the weight of denser tissue such as bone. Sharks lack a gas bladder, but because of the light cartilaginous skeleton and a huge oily liver with a very low specific gravity (the ratio of the weight of a given volume of a substance to that of an equal volume of water), they are only slightly heavier than sea water. In fact, shark liver oil is five to six times more buoyant than sea water. Moreover, certain species have developed particular ways of reducing their weight: for example, the sandtiger shark (*Carcharias taurus*) gulps air into its stomach.

The difference in the density of various sharks is also related to their habitat: pelagic species are less dense than benthic species. Many sharks, particularly the benthic species (and even some pelagic species, under certain conditions), can lie on the sea bottom for long periods, but all sharks have to swim constantly to stay up; if they stop swimming, they sink to the bottom.

SKIN

Shark skin is rough and abrasive. The skin is covered with small 'dermal denticles' or 'placoid scales', usually microscopic in size. A denticle is composed of a pulp, dentine and enamel-like vitrodentine over a bony basal plate or root that is set into the skin. Dermal denticles reduce friction during swimming and noise generated by the shark's movement. Denticles do not increase in size as the shark grows; instead, other denticles are developed. The shape of dermal denticles varies from species to species and from body part to body part, and is also important in the identification of a species.

Fig. 1

Fig. 2

Shark skin is covered by small 'dermal denticles' or 'placoid scales'. The shape of dermal denticles varies from species to species. Fig. 1: gulper shark (*Centrophorus granulosus*); Fig. 2: sandbar shark (*Carcharhinus plumbeus*) (photographs by Luigi Piscitelli).

COLOUR

Colouration is usually grey, brown or blue on dorsal surfaces of the body, and paler or white on ventral surfaces. Almost all sharks have dark backs and white lower surfaces: this colour pattern serves to render the animals almost invisible to their prey at a distance, both from below and from above. Moreover, the white lower surfaces are in shadow, so that the contrast between the upper and lower colourations tends to be neutralised.

The apex and posterior margin of the fins can show white or black colouration, which is especially evident in species such as the oceanic whitetip shark (*Carcharhinus longimanus*) and blacktip reef shark (*Carcharhinus melanopterus*). Some species exhibit light or dark spots on the body, for example the starry smooth-hound (*Mustelus asterias*) or sandtiger shark (*Carcharias taurus*). The more complex colour pattern, displaying spots or stripes, such as in whale sharks (*Rhincodon typus*), in wobbegongs (family Orectolobidae) or in many catsharks (family Schyliorhinidae), is less common. Colouration plays an important role in species recognition.

The zebra shark (*Stegostoma fasciatum*). Almost all sharks have grey, brown or blue backs with white lower surfaces. More complex colour patterns, like those of zebra sharks, are less common (photograph by Alessandro De Maddalena).

REPRODUCTION AND LIFE SPAN

Sharks have a slow rate of growth, and consequently they also have long sexual maturation times. Depending on the species, sharks reach sexual maturity in two to 20 years. The eggs are fertilised internally. The male shark has two claspers, extensions of the pelvic fins that are used to impregnate females. Claspers are formed from the modified inner margin of the pelvic fins. In young sharks the claspers are small, while those in adult individuals are long and calcified. During

A male gray reef shark (*Carcharhinus amblyrhynchos*).
The male shark has two claspers, used to impregnate
females (photograph by Vittorio Gabriotti).

mating, one clasper is inserted into the female cloaca, then the seminal products are carried to the tip of the clasper through a central groove of this organ. In many species, the male bites the female during courtship in order to stimulate the female to copulate; these 'love bites' or mating scars are often evident on the flanks, belly, gill slits, back and fins.

Sharks exhibit one of the following three reproductive methods:

a) oviparous species lay horny egg cases containing embryos nourished by their yolk-sac;
b) aplacental viviparous species produce live young nourished in the uterus by a yolk-sac;
c) placental viviparous species produce live young nourished in the uterus by a placenta formed from a modified yolk-sac attached to the uterine wall.

Some aplacental viviparous species also show oophagy (embryos in the uterus feed on additional unfertilised eggs produced by their mother) and embryophagy (embryos in the uterus feed on their siblings). Aplacental viviparity is the most common reproductive method.

Sharks can have long gestation periods of up to 22 months in the piked dogfish (*Squalus acanthias*). However, the average gestation period is nine to 12 months.

Sharks produce relatively small numbers of young, and litter sizes vary from one in numerous species to 300 in the whale shark (*Rhincodon typus*). Shark pups receive no parental care, but are born fully formed and ready to feed.

A female whitetip reef shark (*Triaenodon obesus*) showing 'love bites' on the gill slit region (photograph by Vittorio Gabriotti)

Many shark species segregate by sex and size. Also nursery areas (areas where only newborns live) have been observed for numerous species. Many sharks reproduce every other year.

Sharks live fairly long lives, having a maximum life span ranging from ten to seventy years or greater, but most species live twenty to thirty years. The highest longevity of all sharks seems to be the piked dogfish (*Squalus acanthias*), which has a maximum life span of at least seventy years.

HABITAT, DISTRIBUTION AND MOVEMENTS

Sharks tolerate a wide range of environmental conditions. Most sharks live on continental and insular shelves and adjacent slopes. Although most large sharks live offshore, they may occasionally venture into shallow water very close to shore, especially in tropical waters and usually at night. Young sharks are often restricted to shallow waters closer to the shores, while adults are commonly wide-ranging.

Sharks can visit the deep parts of the oceans. For example, the Pacific sleeper shark (*Somniosus pacificus*) and the Portuguese shark (*Centroscymnus coelolepis*) hold the record for going deeper than any other shark: the deepest dives recorded for both species are at 1 920-1 930 metres (Portuguese sharks have been caught as deep as 2 719 metres, but there is a possibility that they were captured when the gear was on the way up or down). Presumably most sharks go deeper in the equatorial area. Off West Africa, a shark of unknown species was seen from the bathyscaph FNRS-3 below 3 962 metres.

Some species are active swimmers, while others are bottom-dwellers. Certain sharks also live in freshwater for long periods. The best-known species that has a

A blacktip shark (*Carcharhinus limbatus*) ventures into shallow waters of Walkers Cay, Bahamas (photograph by Harald Baensch).

The blue shark (*Prionace glauca*) is one of the widest-travelling sharks: individuals tagged off England have been recaptured off Brazil (photograph by Walter Heim).

habit of entering rivers and lakes is the bull shark (*Carcharhinus leucas*), which has been found more than 3 000 kilometres upstream from the sea.

Sharks are found in all marine waters of the world. They inhabit tropical, temperate and cold waters, including arctic and antarctic waters. Some sharks are limited in their distribution, but many species are wide-ranging. Abundant species having very wide geographic distribution include piked dogfish (*Squalus acanthias*), blue shark (*Prionace glauca*), oceanic whitetip shark (*Carcharhinus longimanus*), blacktip shark (*Carcharhinus limbatus*), tiger shark (*Galeocerdo cuvier*), bull shark, scalloped hammerhead (*Sphyrna lewini*), basking shark (*Cetorhinus maximus*), shortfin mako (*Isurus oxyrinchus*) and common thresher shark (*Alopias vulpinus*).

Sharks are able to cover great distances, following water currents as the temperatures change. During cold months they are closer to the equator, while in warm months they are further away from it. In tropical areas, the temperature changes are minimal, so that many sharks do not need to migrate large distances. Extensive migrations are also related to feeding and reproduction. The blue shark is one of the widest-traveling sharks, with individuals tagged off England being recaptured off Brazil.

MUTUALISM

Large sharks are often accompanied by remoras and pilot fish. The remoras (*Remora remora*, *Echeneis naucrates* and *Remorina albescens*) are bony fish of the family Echeneidae. Remoras have a dorsal suction disk (formed from their modified dorsal fin) which they use to attach themselves to sharks, mantas, marine turtles and other large creatures, but they use this organ only when the large animal changes direction or slows down. The pilot fish (*Naucrates ductor*) is a bony fish of the family Carangidae often associated with cartilaginous fish, bony fish and marine turtles.

The relationships between pilot fish and sharks, and between remoras and sharks, are cases of mutualism, because both organisms benefit from each of these relationships. Pilot fish and remoras benefit from the relationship with a shark by eating shark scraps of food or excrements and parasites (sharks are hosts to numerous external parasites), as well as by riding the shark's bow wave. Consequently, sharks benefit from the relationship with remoras and pilot fish by being cleaned of parasites.

A live sharksucker (*Echeneis naucrates*) attached to a spotted eagle ray (*Aetobatus narinari*). The relationship between remoras and sharks is a case of mutualism (photograph by Vittorio Gabriotti).

THE ROLE OF THE SENSES
IN FEEDING

Sharks have a highly developed nervous system that illustrates some marvellous physiological structures. When comparing brain to body weight, some sharks have larger brains than many other vertebrates. Their complex brain is necessary to integrate a great deal of sensory information.

The shark's sophisticated sensory system is used to find and catch prey. Odours, sounds and similar low-frequency vibrations, minute electrical currents and visual stimuli can attract a shark. Some of these stimuli are detected from a long distance.

The shark's senses are particularly specialised for locating prey. These cartilaginous fish have four senses: chemoreception, mechanoreception, photoreception and electroreception. Predatory tactics and feeding depend upon several sense organs. Sharks use each of these sense organs as their guide until they reach the prey, but different senses temporarily dominate during the different moments of approaching prey.

OLFACTION

Feeding areas are often located by olfaction. Sharks have a keen sense of smell, which is their most acute sense. They can detect the scent of blood from a distance of about one kilometre; so a bait slick, like a bleeding fish or a whale carcass, will soon attract sharks if they are within one kilometre of the source. Sense of smell enables sharks to locate injured animals and carcasses, such as the large slick formed by a whale carcass which sends out a very strong olfactory signal to sharks swimming in the zone.

The nostrils are situated on the underside of the snout, and are usually partially covered by a nasal flap. This flap separates sea water flowing in from water flowing out. The nostrils lead to an organ called an olfactory bulb, composed of a series of plates of tissue sensitive to chemicals. The olfactory bulb constantly receives a current of water while the shark is swimming. According to neurobiologists Leo S Demski and R Glenn Northcutt, the great white shark has the largest olfactory bulbs of any shark species measured to date.

Sharks follow the scent in different ways: some species criss-cross the odour trail, while others swim upstream against the odour.

Stimulated by blood and food in the water, shark behaviour becomes

Sharks have a keen sense of smell, their most acute sense. Note the numerous small pores located over the head, the external openings of the ampullae of Lorenzini (photograph by Walter Heim).

increasingly aggressive. Under these conditions, many species can exhibit a nervous, savage and dangerous behaviour, which has been commonly observed in whaler sharks (genus *Carcharhinus*), especially when they congregate in large groups. Once sharks get within 250 metres of the prey, they use other senses.

MOVEMENT DETECTION AND HEARING

The lateral line system and the ears enable sharks to detect movements in the sea water. The lateral line is a row of sensory receptors (hair cell receptors) situated along the flanks, extending from the tail to the head, and is pressure-sensitive, enabling the shark to detect water vibrations. Sharks are able to detect both direction and amount of movement in the water from great distances.

A shortfin mako (*Isurus oxyrinchus*). The photograph shows clearly
the lateral line (photograph by Walter Heim).

A wounded creature sends vibrations to the predator that indicate the animal
is in trouble and therefore easy prey. Low-frequency vibrations produced by
wounded prey soon attract sharks if they are within 250 metres of the source.
Sharks are particularly attentive to movements made by injured animals.

Sharks also use this sense to detect sea water currents. They also have two
inner ears, connected to the exterior by narrow canals called endolymphatic
ducts. Hearing is similar to the lateral line system, hair cell receptors being
situated in the inner ear. Hearing in sharks is very sensitive to low-frequency
vibrations like those produced by a wounded prey.

VISION

Prey detection and identification depend heavily on vision. When sharks are
within about 25 metres of prey, they use vision to locate and investigate it. Often
the shark circles the prey while observing it.

The vision of a shark is excellent, because the eyes are highly sensitive. Shark
eyes are similar to those of many other vertebrates. The eyeball encloses the
retina, which is the light-sensitive receptor area. Different parts of the retina
are adapted for bright and dim light, with the consequence that sharks are able
to use their eyes even in low light conditions. The retina contains cone and
rod photoreceptors: cones function in bright light, while rods function in dim

light. The tapetum lucidum is a structure that lies under the retina and reflects incoming light back through the retina to restimulate photoreceptors, thus increasing the sensitivity of the eye. At least some sharks have colour vision.

The eyes of sharks differ in shape and size, and can be very large or small in response to different requirements. Generally, active species have larger eyes and some deep-water species have huge eyes because they need to capture more light. Sharks have immoveable eyelids, but many species have a third eyelid, called the nictitating membrane, formed by an additional fold of the lower eyelid. The nictitating membrane is moveable, and when the shark is feeding the membrane closes over the eye to prevent damage. The great white shark (*Carcharodon carcharias*) lacks this membrane, and in order to reduce the risk of injury it rolls the eye backward when attacking prey.

ELECTRORECEPTION

When sharks are closer to the prey, they can detect the minute electrical currents generated by the prey's nervous system, by using electrical sensors called ampullae of Lorenzini. The ampullae are numerous small organs containing a sensory hair cell filled with an electrically conductive jelly. The external openings of electroreceptors are small pores located over the head, and particularly abundant on the underside of the snout. The long and wide snout provides a perfect position for these electroreceptors.

In some cases detection depends heavily on electroreception. These sophisticated sensors are sensitive to electrical discharges as minute as 0.005 microvolt, and are very useful to find prey buried under sand. Sharks also use this sense to orient themselves using the Earth's magnetic field, particularly for long distance migrations. For this same reason, sharks are also attracted to metals in response to the galvanic currents produced by electrochemical interactions between sea water and metals.

TOUCH AND TASTE

During an attack, even touch and taste provide important contributions. Touch receptors are located over the shark's body, and this sense is used to obtain further information by means of bumping the prey. Taste enables the predator to identify the food before it is ingested: some sharks decide on food palatability while it is lodged in their mouths. Gustatory receptors are located in the mouth and in the pharynx.

Sharks have immoveable eyelids, but many species have a third
eyelid, the nictitating membrane, that is moveable. When the shark
feeds the nictitating membrane closes over the eye to prevent
damage (photograph by Walter Helm).

The great white shark (*Carcharodon carcharias*) does not have the
nictitating membrane, and in order to avoid injury it rolls the eye
backward during an attack (photograph by Vittorio Gabriotti).

A shortfin mako (*Isurus oxyrinchus*) bites a boat propeller. Sharks are attracted to metals in response to the galvanic currents produced by electrochemical interactions between sea water and metals (photograph by Walter Heim).

MOUTH AND TEETH: THEIR FORM AND FUNCTION

Mouth size, tooth shape and jaw morphology are well adapted to the prey that is available to each shark species. Since the mouth leads to the pharynx and gills, feeding and respiration in sharks are closely linked. Some large sharks use the gill system for filter feeding as well as for respiration (see 'Hunting aggregated prey', page 140).

MOUTH

The mouth of most sharks is situated on the undersurface of the head. This characteristic allows a common observer immediately to distinguish most sharks from most bony fish, in which the mouth is terminal rather than ventral. A plausible explanation for the shark mouth position is that these fish may have developed it for feeding on benthic animals. Most shark species diverged to prey on a wide variety of pelagic animals during evolution, but the mouth remained located in the ventral position.

The great white shark (*Carcharodon carcharias*) has large, triangular, highly serrated teeth, the largest teeth of any living shark (photograph by Vittorio Gabriotti).

Mouth form and size vary considerably among species of sharks, and are related to their diet and predatory behaviour. The mouth can be huge, moderately large, or very small; and it can be parabolically curved, slightly curved, or nearly straight. Usually there are upper and lower labial furrows at the corners of the mouth, ranging in size from a few millimetres to a few centimetres.

JAWS

Set of jaws of the great white shark (*Carcharodon carcharias*) (photograph by Alessandro De Maddalena)

Shark jaws are thought to have been derived from a development of the first gill arch. In the most primitive sharks, the cladodonts, the mouth was terminal rather than ventral. These jaws were long, and the upper jaw was fixed tightly to the chondrocranium (braincase). This kind of jaw suspension is called amphistylic, and allowed slight independent jaw movement.

Cladodont descendants were the hybodonts. In these sharks, the jaws shortened, enabling their bite to be more powerful. A significant modification in the jaw structure was the separation of the upper jaw from the chondrocranium. The upper jaw lost the tight connection to the chondrocranium and became loosely suspended from it by ligaments. This made the upper jaw highly mobile, and enabled the shark to protrude its jaws. This kind of jaw suspension is called hyostylic.

A shortfin mako (*Isurus oxyrinchus*) showing
snout elevation and upper jaw protrusion
(photograph by Walter Heim)

Today a few species of sharks have jaws with a modified amphistylic suspension, like the jaws of the frilled shark (*Chlamydoselachus anguineus*). However, most modern sharks have a hyostylic jaw suspension.

The ventral position of the mouth is not an impediment to feeding. Snout elevation and upper jaw protrusion carry the mouth in an almost terminal position. The best example is the great white shark (*Carcharodon carcharias*). Timothy C Tricas and John E McCosker studied the feeding behaviour of the great white shark at Dangerous Reef, South Australia. They discovered that the bite action of a great white shark comprises a sequence of jaw and snout movements. The separate components of the sequence are: a) snout lift, b) lower jaw depression, c) upper jaw protrusion, d) lower jaw elevation, and e) snout drop. Entire duration for a bite action is about 0.9 seconds. During multiple bites the snout remains partially elevated prior to the next bite. The great white shark removes large chunks of prey by biting it and at the same time shaking the head laterally: large individuals can easily remove 20 kilograms of flesh in a single bite. The jaws of large sharks and the associated musculature are extremely powerful. In rare instances, a nurse shark (*Ginglymostoma cirratum*) or a great white shark has bitten a human being and held on so tight that the human had to be pried loose with a tool.

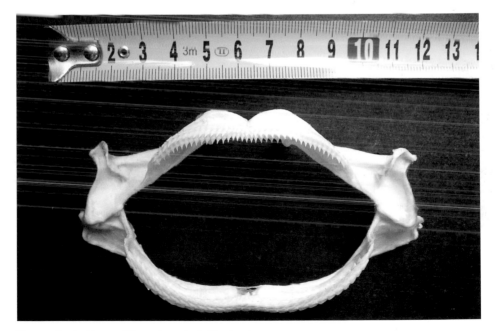

The small set of jaws of the gulper shark (*Centrophorus granulosus*). Depending on species, shark jaws vary considerably in size (photograph by Alessandro De Maddalena).

Jaws vary considerably in size. Some sharks have spectacularly wide jaws, for example the huge whale shark (*Rhincodon typus*), basking shark (*Cetorhinus maximus*), great white shark and tiger shark (*Galeocerdo cuvier*). The author measured a great white shark set of jaws attaining a dried upper jaw perimeter of up to 1.12m. However, the angelsharks (family Squatinidae) have proportionately larger jaws than any other species.

TEETH

The teeth of sharks are merely modified and enlarged dermal denticles, so it is not surprising that the teeth are almost identical in structure to the placoid scales. The teeth are composed of a pulp, dentine and enamel-like vitrodentine over a bony base. Each tooth has a root and a crown, the projection of the crown being called the cusp. The crown has two margins: the lateral and the medial margin.

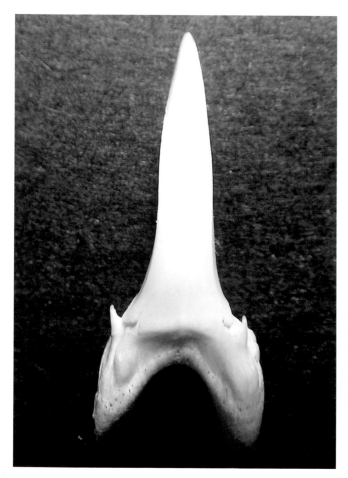

Tooth of the sandtiger shark (*Carcharias taurus*). Each shark tooth has a root and a crown, and the crown has a large main cusp often flanked by one or more cusplets (photograph by Alessandro De Maddalena).

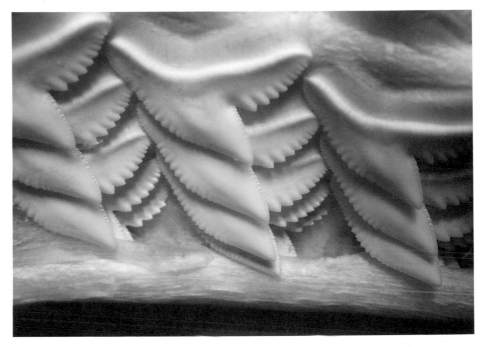

Behind the front teeth there are several rows of replacement teeth.
The picture shows the upper jaw replacement teeth of a tiger shark
(*Galeocerdo cuvier*) (photograph by Alessandro De Maddalena).

The teeth of most primitive sharks are multi-cusped, with a large central cusp and smaller lateral auxiliary cusplets. Still today many species have teeth with a large main cusp flanked by one or more cusplets. A few species, such as the cow shark (family Hexanchidae), have sawlike teeth, showing a cusp and a series of cusplets on each tooth. Usually the tooth anterior face is more flattened than the posterior, which is more curved.

The tooth is not fixed into a socket but is implanted in the connective tissue (tooth bed) of the jaw with the root. Teeth are often broken and quite easily detached (sometimes those of the larger species are found in their prey), but this is not a problem for the shark. The teeth are not permanent, and sharks have a perfect system of regular tooth replacement. Teeth are formed in a groove along the inner jaw surfaces, and behind the front teeth are several parallel rows of replacement teeth. Each jaw typically has five to 15 rows of teeth. As the front teeth are lost, replacement teeth rotate to take their place. The teeth are continuously replaced throughout life. In whaler sharks (genus *Carcharhinus*), each tooth is replaced every eight to 15 days during the first year of life, but in adults replacement slows, and each tooth is probably replaced every month. Depending on the species, sharks may shed 10 000 to 50 000 teeth in a lifetime.

Shark teeth fossilise easily, and for this reason may be the most abundant vertebrate fossils in the world. When fossilised shark teeth were first discovered, they were believed to be tongues of the serpents that Saint Paul had turned to stone while he was in Malta in 59 AD, and consequently they were called 'glossopetrae' or 'tongue stones'.

Some sharks, such as the cookiecutter shark (*Isistius brasiliensis*), swallow their teeth in order to ingest an additional calcium resource. Sharks with small teeth usually have more than one functional row in the jaws, while species with large teeth usually have one or two functional rows.

The number of front teeth in the upper and lower jaws is constant, and is used to identify species. In order to represent the number of teeth in a shark mouth as a dental formula, the teeth in the first row of both jaws are counted. For example, the great white shark dental formula is usually 13-13 / 11-11, meaning that the great white shark usually has 13 teeth in each quadrant of the upper jaw and 11 teeth in each quadrant of the lower jaw. We refer to four quadrants when describing teeth, so the dental formula is read: left side of the upper jaw - right side of the upper jaw / left side of the lower jaw - right side of the lower jaw. To arrive at the total number of teeth in the outer row of upper and lower jaws, sum the numbers: 13 + 13 + 11 + 11 = 48.

The dental formula often shows a small variability. For example, the great white shark formula has variability 12 to 14-12 to 14 / 10 to 13-10 to 13. The blue shark dental formula is usually 15-1-15 / 14-1-14, meaning that blue sharks usually have 15 teeth in each quadrant of their upper jaw, 14 teeth in each quadrant of their lower jaw, and one tooth at the symphysis, between the right and left side of each jaw.

Shark teeth are highly specialised. The teeth exhibit a wide variety of shapes, since the shape varies between species according to what they eat and the lifestyle they lead; in fact, teeth are perfectly adapted to size, form and structure of prey, so tooth adaptations enable sharks to feed efficiently. Teeth are therefore an invaluable means of identification; in some cases a single tooth is sufficient to identify the shark.

There are three main tooth shapes common to sharks with similar feeding ecologies, and numerous variations.

a) Teeth adapted for sawing or shearing pieces from large animals such as sharks and marine mammals: these teeth are large, triangular, sharp, and with or without serrate edges like those of the great white shark and tiger shark.

b) Teeth adapted for seizing smaller fast prey such as small schooling fish: these teeth are narrow and curved, and tend to be moderately to very long; for example, the shortfin mako (*Isurus oxyrinchus*), salmon shark (*Lamna ditropis*) and sandtiger shark (*Carcharias taurus*).

c) Teeth adapted for crushing hard prey such as molluscs and crustaceans: these teeth are smooth or arranged in a pavement formation, such as those found in the smooth-hounds (*Mustelus* spp.).

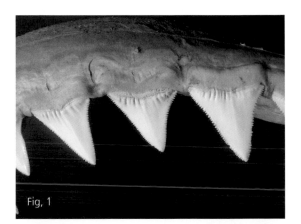

Fig. 1

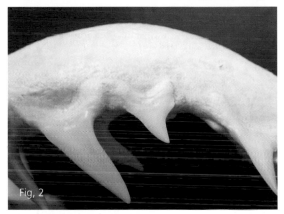

Fig. 2

Fig. 3

The three main tooth shapes: Fig. 1: large, triangular, sharp teeth adapted for sawing pieces from large prey (great white shark, *Carcharodon carcharias*, upper jaw), Fig. 2: narrow, long and curved teeth adapted for seizing smaller fast prey (shortfin mako, *Isurus oxyrinchus*, upper jaw); Fig. 3: pavement-like teeth with a low cusp adapted for crushing hard prey (smooth-hound, *Mustelus mustelus*, lower jaw) (photographs by Alessandro De Maddalena).

Teeth have been modified in a number of ways. Their wide diversity of uses is reflected in their morphology. In most species, the teeth of the upper jaw are very different in shape from those of the lower jaw. Teeth in the lower jaw are often smaller and narrower than those in the upper jaw. Often the lower teeth serve to stab into and hold the prey securely while the upper teeth serve to cut off a piece of meat. Anterior teeth are larger than the teeth that follow them towards the mouth corner.

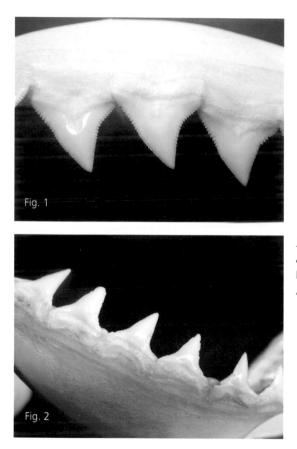

Java shark (*Carcharhinus amboinensis*): upper jaw (Fig. 1) and lower jaw (Fig. 2) (photographs by Alessandro De Maddalena)

In most sharks, the teeth are usually of a similar shape around the upper or lower jaws, but some sharks have two different types of teeth: blunt crushing teeth posteriorly, and sharp teeth anteriorly. For example, the sharks of the family Heterodontidae are so named for having 'different teeth' around their jaws. Many species show a small or conspicuous toothless gap at the symphysis, while others have smaller teeth in this area. Usually teeth can be seen only when the shark opens its mouth, but in a few species the lower teeth protrude prominently and always remain visible.

The great white shark has large, triangular, highly serrated teeth, ideal for cutting up hard bones of marine mammals and large fish such as adult bluefin tuna (*Thunnus thynnus*), swordfish (*Xiphias gladius*) and other sharks. The teeth in the upper jaw are broader than those in the lower jaw. The bluntnose sixgill shark (*Hexanchus griseus*) has small and narrow upper teeth with a curved cusp and 0 to 4 small cusplets, and large sawlike lower teeth with a cusp and 4 to 7 cusplets gradually decreasing in height (in the adult the cusp becomes larger, while in the young it is only slightly larger than the first cusplet). The tooth at the symphysis is very different in shape, being narrower and having a cusp without cusplets, with strongly serrated margins. The tiger shark has large, flat, highly serrated, cockscomb-shaped teeth, with a large notch on the lateral margin, ideal for cutting up marine turtles. The large serrae are secondarily serrated. The blue shark (*Prionace glauca*) has broad, long, strongly serrated, curved, oblique upper teeth, without cusplets, and the lower teeth are smaller, long, narrow, pointed and oblique, and have cutting edges.

Teeth vary considerably in size. Some species have spectacularly enlarged teeth, like the great white shark, longfin mako (*Isurus paucus*) and shortfin mako. The largest teeth belong to the great white shark, which can measure

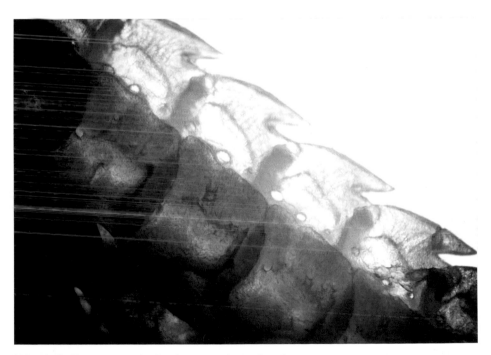

Velvet belly (*Etmopterus spinax*): microscope photo of teeth.
Depending on species, shark teeth vary considerably in size
(photograph by Luigi Piscitelli)

Megatooth shark (*Carcharodon megalodon*): fossil
tooth (photograph by Alessandro De Maddalena)

6.4cm (total height) and 5.1cm (enamel height). The largetooth cookiecutter
shark (*Isistius plutodus*), which measures up to 42cm, has proportionately larger
teeth than any other species.

Body length estimates of large sharks based on examination of the teeth and
jaws is often a source of error and exaggeration. Consequently, the size of the
shark must be evaluated cautiously.

The largest fossilised teeth are those of the extinct megatooth shark
(*Carcharodon megalodon*), recorded in the late Cenozoic. Based on the largest

available teeth, that is 16.8cm in total height, palaeontologists estimated a maximum total length of 15.9m for the megatooth shark. This immense animal was big enough to eat the largest prehistoric sea creatures. Megatooth shark bite scars are sometimes found on the fossil remains of cetaceans.

Tooth shape is also related to shark age. Tooth shape alters as a shark grows larger and feeds on different animals. In order for the shortfin mako to eat fast pelagic fish, it is born with narrow teeth; but as it grows, these become thick and strong to accommodate larger prey such as swordfish and small cetaceans. Young specimens are sometimes mistaken for other shark species because of their tooth shape; for example, young porbeagles (*Lamna nasus*) are sometimes mistaken for shortfin makos because their teeth lack cusplets; and young white sharks are mistaken for porbeagles because their teeth partially lack serrated edges and show cusplets. We discuss this topic again in 'Changes in diet', page 74.

In addition to feeding, teeth are also used as weapons in defence against predators (such as other sharks) and in fighting individuals of the same species. In many species, male sharks have longer teeth than females. In fact, as stated previously, male sharks commonly use their teeth for grasping the female body before and during copulation ('love bites').

THE DIGESTIVE SYSTEM

As in other vertebrates, digestion in sharks takes place in the buccal cavity, pharynx, oesophagus, stomach and intestine. The mouth, pharynx and oesophagus are sufficiently wide to enable the animals to swallow large food items. The oesophagus contains numerous internal fingerlike projections called papillae.

Following the oesophagus, the digestive system continues with the stomach. Here the food is acted on by enzymes. The stomach has longitudinal folds or rugae. Shark stomachs are 'U' or 'J' shaped, consisting of two portions: the cardiac stomach and the pyloric stomach. The cardiac stomach is very large and saclike, while the pyloric stomach is narrower. The stomach capacity of the shortfin mako (*Isurus oxyrinchus*) is 10% of the body weight. The shark stomach is large in order to enable these formidable predators to ingest whole animals, large chunks of prey or a large amount of smaller prey. Consequently, being able to ingest large amounts of food at a time they do not need to feed often. Moreover, food can be stored in the cardiac stomach for long periods of time. Often when a shark is eviscerated, prey is found in a near-perfect state of preservation.

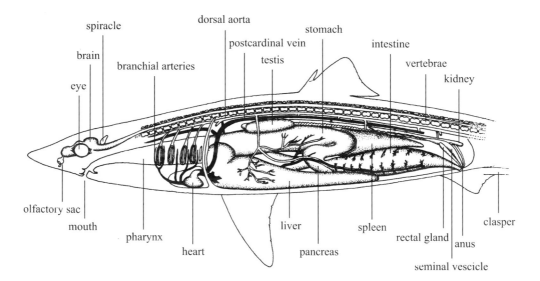

A piked dogfish (*Squalus acanthias*) showing the shark anatomy and digestive system (drawing by Alessandro De Maddalena, redrawn after Storer and Usinger, 1965)

The great barracuda (*Sphyraena barracuda*). Sharks eat even these formidable predators: an 87cm barracuda has been found in a great white shark stomach (photograph by Vittorio Gabriotti).

For example, Daniel W Gotshall and Tom Jow reported the case of a 3.66m Pacific sleeper shark (*Somniosus pacificus*) that, when caught, regurgitated about 136 kilograms of fish, mostly rex sole (*Glyptocephalus zachirus*) and Dover sole (*Microstomus pacificus*), but also three chinook salmon (*Oncorhynchus tshawytscha*) weighing 1.3 to 2.2kg each, and a 4.5kg Pacific halibut (*Hippoglossus stenolepis*). In another recorded instance, a great white shark (*Carcharodon carcharias*) had 150 dungeness crabs (*Cancer magister*) and red rock crabs (*Cancer productus*) in its stomach.

Often ingested prey is found intact or marked by only a few superficial teeth marks. A shark can swallow whole animals attaining a total length of up to 40% of its own size. A spinner shark (*Carcharhinus brevipinna*) caught off New South Wales, Australia, contained a whole tuna. An almost intact 54kg swordfish was found inside a shortfin mako. A 2.53m great hammerhead (*Sphyrna mokarran*) caught off KwaZulu-Natal, South Africa, contained a whole 86cm dusky shark (*Carcharhinus obscurus*).

Sometimes very large animals, such as tunas, seals, marine turtles and other animals (including men in armour), have been found basically intact in great white shark stomachs. Researcher Geremy Cliff and colleagues examined great white sharks that ingested an 87cm barracuda (*Sphyraena* sp.) and a 95cm dusky shark (the length of this shark was measured as precaudal length, that is noticeably shorter than the total length). A 5m specimen caught near Los

Angeles, California, contained two whole sea lions weighing 78.7 and 56.2kg. The stomach of an estimated 6.68-6.81m great white shark, caught off Filfla, Malta, contained a whole 2.2m blue shark (*Prionace glauca*), a 60cm loggerhead sea turtle (*Caretta caretta*) and a 2.5m dolphin in two or three parts. A 4.72m specimen caught in Florida contained two whole sandbar sharks (*Carcharhinus plumbeus*) measuring between 1.8 and 2.1m in length. Sixteenth-century French naturalist, Guillaume Rondelet, suggested that the great fish that swallowed the prophet Jonah in the Biblical account was a great white shark!

Sharks are able to evert the stomach, possibly in order to provide a means to empty the stomach of indigestible objects. Aquarists have observed captive sharks evert their stomach and behave as if nothing has happened. The stomachs of these individuals are usually returned to the interior of their body cavity, and the sharks remain in good health. This behaviour has often been observed when sharks are captured. Of 128 blue sharks examined by Australian researcher John D Stevens, 77 had everted stomachs. A high percentage of everted stomachs has also been observed in the oceanic whitetip shark (*Carcharhinus longimanus*), silky shark (*Carcharhinus falciformis*) and dusky shark. However, this behaviour is less common in other species such as the shortfin mako, great white shark, bull shark (*Carcharhinus leucas*) and smooth hammerhead (*Sphyrna zygaena*).

After the stomach, the digestive system continues with the intestine. The shark intestine is relatively short. The anterior part is the duodenum. The duodenum, pancreas and liver are closely linked. The pancreas is an important digestive gland, which secretes digestive enzymes. Pancreatic secretions flow out of the pancreas through a duct into the duodenum.

The liver (see also 'Buoyancy', page 16 and 'Rate of food consumption', page 46) is a huge organ (it may comprise as much as 25% of the total body weight), that consists of two lobes. Liver cells secrete bile that emulsifies lipids. The bile is stored in a sac called the gall bladder, and is secreted into the duodenum through a duct.

The duodenum leads to the most important portion of the intestine, the ileum, that contains an interesting structure called the spiral valve. The spiral valve is a corkscrew-shaped internal structure, and actually is not a valve: it serves to increase the absorptive surface of the intestine without increasing its external size. This adaptation provides sharks with the necessary space for a very large liver and stomach. The absorption of the end products of digestion occurs in this section. The spiral valve shape differs among species, and in some sharks this portion is rolled up on itself in scroll fashion.

The ileum leads to the last section of the intestine, the rectum. The rectal gland is located in this section which removes excess salt from the blood and opens by a duct into the rectum. The digestive system terminates with the anus, that opens into the cloaca. In this small chamber, located on the ventral side of the animal between the two pelvic fin bases, the urinary and genital tracts also open to the outside.

The morphology of the digestive system varies from species to species, and in a few cases is of importance in the identification of a species. For example, the number of spiral valve turns and the duodenum length enable a researcher to distinguish the longfin catshark (*Apristurus herklotsi*) from the longhead catshark (*Apristurus longicephalus*).

Studies have shown that shark digestion is slow compared to that of bony fish. In fact, initial digestion of the food is relatively fast, taking about 24 hours, but it takes 1.5 to 5 days for the food to be completely dissolved, the nutrients totally absorbed, and the waste products completely excreted. However, the rate at which food is digested is closely related to the activity level of each species, the shark body temperature (when the body is warmer digestion may be faster), and the sea water temperature (when the sea water is colder digestion may take longer). The benthic nurse sharks (*Ginglymostoma cirratum*) digest their meals in six days. The pelagic blue shark takes at least three days to digest an average-sized meal. One of the most active pelagic sharks, the shortfin mako, digests its food in 1.5-2 days. This will be further discussed in the next section.

The blue shark (*Prionace glauca*) takes at least three days to digest an average-sized meal (photograph by Walter Heim).

Rate of food consumption

Many people think of sharks as voracious monsters, continuously eating, but they actually consume a small amount of food. The shark's average meal is 3% to 5% of its body weight. Sharks feed intensively for a short time, and then feed very little for a longer period of time, so they may often pass up potential prey. Most sharks feed at one- or two-day intervals. The lemon shark (*Negaprion brevirostris*) feeds actively for about eleven hours, and then slowly for the next 32 hours. There is a short period of feeding activity when sharks are in a feeding mode, and a longer period of digestion when feeding is minimal; most sharks feed infrequently because cold-blooded animals have a much slower metabolism than warm-blooded animals. Sharks are also able to stop feeding for several weeks. During this time they live off reserves in the large liver that is particularly rich in oil.

An Australian fur seal (*Arctocephalus pusillus doriferus*) resting on a rookery. It has been estimated that about thirty kilograms of fatty tissue taken from a pinniped would provide the energy necessary to sustain a 4.6m great white shark (*Carcharodon carcharias*) for 1.5 months (photograph by Vittorio Gabriotti).

Even the great white shark (*Carcharodon carcharias*) feeds infrequently. In a study conducted at Año Nuevo Island, California, USA, only two probable feeding bouts on pinnipeds occurred during a 15-day period for which five sharks were monitored, one during night-time and another during daytime. Carey and colleagues estimated that a 4.6m great white shark could survive 1.5 months between meals. It has been estimated that about thirty kilograms of fatty tissue taken from a whale or a pinniped would provide the energy necessary to sustain a great white shark for this period. In fact, great white sharks, similarly to blue sharks (*Prionace glauca*), seem to feed selectively on the energy-rich blubber layer of a cetacean carcass. A juvenile elephant seal (*Mirounga* sp.) contains more energy than required for sustenance of a single white shark, since a 140kg elephant seal would have about 67kg of fatty tissue, which would provide twice the energy necessary to sustain a white shark for 1.5 months. When a great white shark captures a pinniped, other white sharks often arrive at the site of the kill to feed.

A large cetacean carcass contains enough energy to sustain numerous great white sharks for long periods. The nutritional benefit received by sharks from a whole whale is enormous; consequently whale carcasses are a typical food source for many shark species. It appears that sharks must expend considerable time and energy looking for prey and in hunting activity, and so it is beneficial for a large shark to consume one large food item rather than to capture numerous small items.

Rate of consumption is estimated based primarily on the amount of food found in shark stomachs. Many captured sharks have empty stomachs, having probably eaten at least 24 hours before they were captured (however, we must consider that many species are able to evert their stomachs). For example, of 26 spinner sharks (*Carcharhinus brevipinna*) examined by John D Stevens, 18 had empty stomachs, and of 139 frilled sharks (*Chlamydoselachus anguineus*) examined by Tadashi Kubota and colleagues, 102 had empty stomachs.

Compared to similar-sized bony fish, sharks do not require large amounts of food: they do not eat as often, and digest more slowly. While many bony fish eat an equivalent of their body weight in a few days, sharks can consume the equivalent of their body weight in one to 16 months. The number of different food items found in the stomach is usually low, indicating that sharks generally do not consume additional prey after they have already eaten. Shark stomach contents amount to an average of one-fourth their stomach capacity.

There are also many reports of sharks with full stomachs being caught, but generally these cases represent a small percentage. Even though these individuals had just eaten, they were still attracted to bait.

Food consumption needs to compensate for energy expended in growth and activities including normal swimming, prey search, predation and reproduction. Obviously, food consumption is related to the activity level of a particular species, which also relates to the rate at which food is digested. Consequently, the daily ration or food consumption per day is different for each species. A sedentary shark such as the nurse shark (*Ginglymostoma cirratum*) requires a small amount of food, so it is not surprising that it eats just 0.2% to 0.3% of its body weight per day. Pelagic predators, such as the blue shark and sandbar shark (*Carcharhinus plumbeus*), eat 0.2% to 0.6% of their body weight per day. Young lemon sharks (*Negaprion brevirostris*) consume 1.7% of their body weight per day. The fast-swimming shortfin mako (*Isurus oxyrinchus*) consumes 3% of its body weight per day.

A sedentary shark such as the nurse shark (*Ginglymostoma cirratum*) requires a small amount of food (photograph by Harald Baensch).

Sharks are difficult to keep in captivity. Under these conditions they consume 0.4% to 2% of their body weight per day, but can also stop feeding for many weeks or months, and die. Aquarists spend considerable time trying to get the sharks to take food, but the animals refuse everything they are offered.

Some interesting observations have been made by Monterey Bay Aquarium curator David C Powell. Pelagic sharks need to swim actively to obtain the amount of oxygen required by their metabolism, and species such as the blue shark have evolved physiological characteristics for long-distance swimming in a relatively straight line. In natural conditions these species eat a small amount of food because they do not require much energy. If these sharks are captured and forced to stay in a tank where they have to be continuously turning, they consume much more energy, because the costs of turning and acceleration are significantly higher than cruising in a straight line. As stated previously, the large shark liver is an energy reserve and also serves as a buoyancy organ. As the lipids in the liver are used to obtain additional energy, the shark loses weight, and losing the oil in its liver causes its specific weight to increase. Consequently, the animal has to swim harder to stay up and begins to exhibit an abnormal swimming posture. The captive shark consumes its energy reserves faster than its metabolism replaces them, and finally it dies. The liver size is a good index of shark health, and those that are in good condition have larger livers.

Oxygen consumption can be measured using respirometry to quantify the shark's energy requirements. Oxygen consumption increases with increases in activity, such as swimming speed and tailbeat frequency. Therefore, these characteristics can be used as predictors of the metabolic rate of sharks.

The large amount of food required by the fast-swimming shortfin mako is not only for physiological maintenance and movement, but also to accommodate a very rapid growth rate. Fisheries biologist Chuck Stillwell studied the food consumption of the shortfin mako. Newborn makos have a birthweight of 2.3-2.7kg. In the first three years of life, their weight may increase an average of 27.2kg per year. Males reach about 136 kilograms in 4.5-5 years, while females grow to about 226 kilograms over seven years. In contrast to the shortfin mako, a pelagic but relatively less active species such as the sandbar shark (*Carcharhinus plumbeus*) can reach about 41-45kg in 12-14 years. Why do shortfin makos have this particular physiology?

While most sharks have a body temperature equal to that of the surrounding sea water, some species of the order Lamniformes exhibit regional endothermy, and maintain a higher body temperature than the sea water. These sharks are warm-bodied because they have heat-retaining systems. These species are the shortfin mako, great white shark, porbeagle (*Lamna nasus*), salmon shark (*Lamna ditropis*) and thresher sharks (*Alopias* sp.). Most sharks of the family Lamnidae have an elevated body temperature, from 2°C to 14°C above the sea water temperature (depending on species and body organs, but generally higher temperatures exist around the digestive system and muscles).

Red muscles are the most powerful during swimming. Endothermic sharks

The shortfin mako (*Isurus oxyrinchus*) is a warm-bodied animal.
While most sharks have a body temperature equal to that of the
surrounding sea water, some species of the order Lamniformes
maintain a higher body temperature than the sea water;
consequently they are powerful, fast and agile, and can make
excursions into cold water (photograph by Walter Heim).

have large amounts of red muscle tissue located deep in the trunk, close to the vertebral column (these muscles form the axial musculature), while other species have these muscles more superficially located. The red muscle tissue is connected to the circulatory system by a complex network of arteries and veins called 'rete mirabile' ('marvelous net') that acts as a countercurrent heat exchanger to reduce heat loss to the body periphery. Heat generated in the red muscles by swimming warms the blood, which passes through the venules in the rete mirabile, and the heat is transfered to the parallel counter-flowing arteries. Thus heat is conserved in the shark body, rather than dissipating to the environment, and elevates the muscular and visceral temperature.

Heat is a form of energy, and consequently warm-bodied sharks have more energy at their disposal. An increase in temperature results in an increase in the muscular efficiency; the rise in temperature increases the rate of chemical reactions, resulting in an increased muscle contraction-relaxation rate. The food ingested is digested and assimilated at an accelerated rate, providing the metabolic energy required by these large predators. This adaptation enables them to swim at high speeds for extended periods, and also allows excursions into cold water.

Therefore, it is not surprising that warm-bodied sharks are powerful, fast and agile. The shortfin mako may be the fastest shark, with maximum speeds of 35-56kmh. Adult shortfin makos, common thresher sharks, great white sharks, salmon sharks and porbeagles are known for leaping from the water in high breaches that attain an elevation of several metres.

A similar modified circulatory system is also present in some bony fish, such as the fast tunas and billfish.

The metabolism of these very active sharks requires a large amout of oxygen. Therefore, the gill slits are larger in warm-bodied species. These sharks are particularly difficult to keep in captivity because they require fast movement to stay alive. Forcing them to stay in a relatively large tank is insufficient. Only 6% of shortfin makos and 16% of great white sharks were alive when found in the protective nets off KwaZulu-Natal, South Africa; this low percentage reflects their higher oxygen requirements.

WHEN SHARKS FEED

Certain shark species are most active at night or around dawn and dusk; others appear more active during the day. However, most sharks have crepuscular or nocturnal feeding habits, and mainly feed at twilight or in darkness. The hunting success of many sharks depends on the element of surprise, and they take advantage of the reduced ambient light levels to attack their prey. The Natal Sharks Board, South Africa, reported that during the years 1966 to 1972 over 95% of the sharks captured became trapped in the nets between dusk and dawn.

Nocturnal species include the Pacific angelshark (*Squatina californica*), horn shark (*Heterodontus francisci*), Port Jackson shark (*Heterodontus portus-jacksoni*), tawny nurse shark (*Nebrius ferrugineus*), swellshark (*Cephaloscyllium ventriosum*), epaulette shark (*Hemiscyllium ocellatum*), whitetip reef shark (*Triaenodon obesus*), grey reef shark (*Carcharhinus amblyrhynchos*), blacktip reef shark (*Carcharhinus melanopterus*), tiger shark (*Galeocerdo cuvier*),

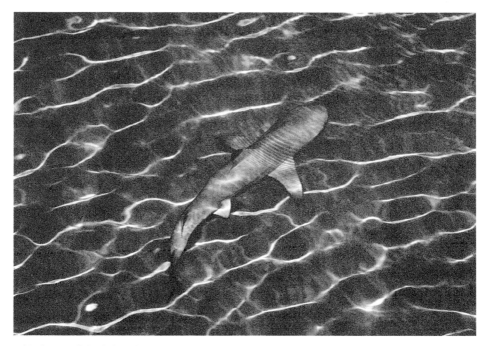

A blacktip reef shark (*Carcharhinus melanopterus*) swims in the night in very shallow waters. Many shark species increase foraging activities during the night (photograph by Vittorio Gabriotti).

lemon shark (*Negaprion brevirostris*), blue shark (*Prionace glauca*), scalloped hammerhead (*Sphyrna lewini*), small-spotted catshark (*Scyliorhinus canicula*) and bluntnose sixgill shark (*Hexanchus griseus*). Almost all requiem sharks (family Carcharhinidae) are considered to be nocturnal.

The bonnethead shark (*Sphyrna tiburo*) is a diurnal species, while the great white shark (*Carcharodon carcharias*) is equally active during the day and night.

Some nocturnal hunters rest in crevices and caves during the day. This behaviour is often observed on coral reefs, which support a wide range of sharks, both nocturnal and diurnal. One of the most well-known species that rests in caves is the whitetip reef shark. Large aggregations of these animals have often been observed in caves and passes located in coral reefs. As the daylight fades, numerous animals try to find somewhere safe to hide away from nocturnal hunters. Then the sharks that rest in crevices and caves during the day leave their refuges to begin their night patrol.

Daytime feeding by nocturnal species is often reported, as these sharks readily respond to feeding stimuli and eat when occasion is present during the day, but hunting increases at night.

American researcher, A Peter Klimley, and colleagues have studied diel movements of scalloped hammerhead sharks near the seamount El Bajo Espiritu Santo, in the Gulf of California, Mexico. This site attracts numerous scalloped hammerheads as well as large pelagic bony fish. The hammerheads remain grouped at El Bajo Espiritu Santo during the day, depart prior to dusk, move separately into the surrounding area at night, and return to their home at the seamount at dawn. The scalloped hammerheads do not feed while schooling at El Bajo Espiritu Santo during the day, and feed on fish and cephalopods when swimming separately into the surrounding waters at night. Similar to scalloped hammerhead sharks, grey reef sharks remain grouped during the day and move separately to catch their prey at night.

Until a few years ago, it was believed that great white sharks were primarily daytime hunters. These animals were observed to attack pinnipeds during the day at the South Farallon Islands, off San Francisco, California, USA. Researcher Wesley Rocky Strong Jr observed a lower rate of white shark attraction at night when studying these animals near sea lion colonies in Spencer Gulf, South Australia. Moreover this species has colour vision, consistent with daytime predation.

Peter Klimley and colleagues studied great white shark predation on pinnipeds at Año Nuevo Island, California, USA. The monitored sharks visited the area near the seal colony each day. The sharks patrolled Año Nuevo Island during both daytime and night-time, and they spent an equal amount of time and activity near the seal colony at all times of the day. The sharks tracked at this site fed on pinnipeds two times, once at night-time and once during daytime.

Scalloped hammerheads (*Sphyrna lewini*). In the Gulf of California, Mexico, these sharks remain grouped and do not feed during the day, but catch fish and cephalopods separately at night (photograph by Harald Baensch).

Certain species show diel variations in their vertical distribution, such that they swim in deep zones during the day and migrate from deep waters to the surface at night. The megamouth shark (*Megachasma pelagios*) spends the daytime in deep waters and ascends to midwater depths at night. This vertical migration may be a response to the movements of the plankton on which megamouth sharks feed. Another vertical migrator is the velvet belly (*Etmopterus spinax*). This small shark lives close to the sea bottom during the day and ascends toward the surface to feed on small fish and squid at night. Another predator that ascends to the surface at night to feed is the bluntnose sixgill shark (*Hexanchus griseus*).

These diel vertical migrations can be enormous: researchers have hypothesised that the pygmy shark (*Euprotomicrus bispinatus*) may ascend in excess of 1 500m to the surface at night.

Tidal height influences predation by sharks on their prey. Certain species exhibit the highest foraging activity during certain states of the tidal cycle. Some species, such as the sandbar shark (*Carcharhinus plumbeus*), move inshore to feed at high tide. On the Great Barrier Reef, Australia, blacktip reef sharks and sicklefin lemon sharks (*Negaprion acutidens*) move into intertidal waters to catch their prey at flood tide. Predation by the great white shark on pinnipeds at South Farallon Islands, California, USA, is more frequent during higher tides than during lower tides, because northern elephant seals (*Mirounga angustirostris*) are more numerous in the water during higher tides. The South Farallon Islands pinniped population has reached carrying capacity, and space at high tide is scarce. Consequently, during higher tides, juvenile pinnipeds are forced to enter the water owing to lack of space.

Peter Pyle and colleagues studied the effects of environmental factors on the feeding behaviour of great white sharks at the South Farallon Islands, and discovered that other factors, including water clarity, weather, lunar illumination and sea temperature, may affect frequency of predatory activities.

THE STUDY OF SHARK FEEDING BIOLOGY

The lack of data on the feeding biology of many shark species is surprising, given their wide distribution and large size; but sharks are difficult to study, and many gaps exist in our knowledge of these fish. The study of shark feeding biology requires knowledge of sharks and their prey, time and patience; and the number of scientists working on the predatory tactics of sharks has always been relatively small.

Among the many difficulties encountered when studying shark feeding biology are the following main problems. Sharks are difficult to study in

Divers observing a great white shark (*Carcharodon carcharias*). Sharks are difficult to study in the wild: they can be dangerous or potentially dangerous, uncommon or rare, elusive and timid, and they usually catch prey without showing at the surface (photograph by Vittorio Gabriotti).

the wild. Some of the most interesting species are dangerous or potentially dangerous. Numerous species are uncommon or rare, and many are infrequently encountered. Most sharks are elusive and timid, usually avoiding people and catching prey without showing at the surface. Many species have crepuscular or nocturnal feeding habits, which makes observation by researchers difficult. The feeding process of many species has never been observed. Most sharks are eviscerated by fishermen immediately upon capture, and consequently it is difficult to collect data on their diet.

Another problem is that most studies on feeding biology cannot be done in captivity. Most species are difficult to keep in captivity, requiring careful capture and transport, large tanks, high water purity; often they refuse to feed, and their dietary needs are poorly known or unknown. Some species that live close to shore adapt to life in captivity, but those that live in open water often die within a few days.

Finally, field research is usually expensive. In many countries, studies on sharks are almost totally neglected by governments.

The primary objective for studying shark feeding biology is to document predatory behaviour and diet. Much of our fragmentary knowledge about the feeding behaviour of sharks is based on observations of only a few species. However, our knowledge is noticeably greater than what was available just a few years ago. A great deal about the feeding biology of sharks is learned by studying these creatures in different ways. For example, researchers study jaw and teeth anatomy, neuro-muscular physiology, and bite actions using photographic and cinematographic techniques; shark sensory analysis takes place in laboratories, and scientists observe approach behaviour to baits, and try to determine shark reactions to various stimuli. Through the use of scientific instrumentation our knowledge has grown significantly.

Much information on the feeding biology of sharks is obtained from commercial fishermen, and anecdotal accounts identifying prey items of many sharks are scattered throughout the literature. But only a few comprehensive studies of food habits have been undertaken. Understanding the behaviour and biology of shark prey can provide important insights into understanding sharks' predatory tactics.

Direct evidence of what a shark eats may usually be obtained by three methods:
a) examination of shark stomach contents through dissection of dead specimens;
b) direct observations of shark predation and scavenging;
c) examination of signs of shark predation and scavenging on carcasses and wounded animals.

The purpose of the following paragraphs is to describe these different methods.

EXAMINATION OF STOMACH CONTENTS

Examination of shark stomach contents is the primary method for studying the shark diet. Many different sampling methods are used: material is available from research cruises made aboard research vessels, and even sport fishing tournaments can provide a valuable source of samples. However, commercial fisheries provide the most important source of material. Numerous sharks are available at fish markets, but data on shark stomach contents are often hard to collect owing to the fact that specimens are generally eviscerated by fishermen immediately upon capture, and the viscera dumped at sea. This problem is commonly encountered in the case of large and medium-sized sharks.

So, shark stomachs are often excised at sea by fishery observers for later analysis in the laboratory. The oesophageal and pyloric ends of the stomachs are secured with plastic ties, and the samples are bagged, labeled and frozen. Data on set and haul time, location, water depth, water temperature, shark size, weight, sex and maturity state are recorded. Samples are collected from different shark size groups, and the stomachs are then processed in a laboratory. Researchers weigh the stomach contents, estimate the degree of food digestion, and try to identify the prey (it is often impossible to identify the species positively, so the contents are identified to the lowest possible taxon). They also perform a series of statistical analyses, and compare diets between size and age classes, locations and seasons. Arbitrary size-classes of sharks are used to investigate changes in the diet associated with growth. Records of the stomach contents of sharks caught worldwide are combined and analysed on a global scale.

A common problem encountered examining shark stomach contents is that when prey is found in a shark stomach, no-one knows whether the animal was alive or dead when ingested. A fundamental distinction must be made between predation on a live animal and scavenging a dead individual. It is often impossible to find indications that the stomach contents were taken through active predation rather than post-mortem scavenging.

DIRECT OBSERVATION OF PREDATION AND SCAVENGING

The observation of shark predation and scavenging is the fundamental method for studying shark predatory tactics; studies on predation must be long-term in nature.

Direct observation of shark predation is possible when it takes place at the sea surface; sometimes sea birds converging over a given area can be reliable indicators of a successful shark attack. Another sign of a predatory attempt can

Sharks are attracted to research vessels by huge quantities of macerated fish, horse meat and blood used as bait (photograph by Roberta Gabriotti).

be a great splash at the sea surface.

Some sharks are more likely to be seen at the surface than other species; for example, this occurs with tiger sharks (*Galeocerdo cuvier*) feeding on albatross near the Hawaiian Islands; or with great white sharks (*Carcharodon carcharias*) preying on pinnipeds in Spencer Gulf, Australia, in False Bay and near Dyer Island, South Africa, and at the Farallon Islands and Año Nuevo Island, California, USA, since these sites are home to colonies of seals and sea lions. These locations are the most famous white shark hunting grounds in the world, and researchers come from many countries to observe shark feeding behaviour. These locations provide ichthyologists with a fundamental platform of opportunity to study shark/prey interactions.

Other direct observations of shark feeding happen when sharks scavenge on

Researchers often use cages to study the behaviour of
dangerous sharks (photograph by Vittorio Gabriotti).

a carcass, such as a dead whale. Direct observations of shark predation on aggregated prey is sometimes possible when it takes place at the sea surface, such as in the case of blue sharks (*Prionace glauca*) feeding on cephalopods, or bronze whaler sharks (*Carcharhinus brachyurus*) attacking schools of pilchard.

Often, the eyewitnesses to shark predation are not researchers but occasional observers, such as scuba divers, bathers and fishermen. This was the case when tourists observed a tiger shark of approximately 2m attacking and killing a newborn dolphin in Shark Bay, Western Australia.

In order to observe sharks, it is important to establish a food source to which the animals are attracted. This enables researchers to observe, from the boat and from cages, sharks feeding on baits both at and below the sea surface. Sharks are often attracted to the research vessels by the use of huge quantities of macerated fish, tuna, horse meat and blood as bait. This mixture of macerated fish, blood, oil and other attractants is called 'chum'.

Sometimes it takes weeks for a shark to arrive. The shark cages are constructed of aluminum or steel, are securely attached to the boat, and are usually located at the surface or at a depth of a few metres. Cages have viewing ports that allow for photographing and filming. Nevertheless, attempting to attract dangerous sharks is prohibited in many areas; so it is necessary to check with the local authority having jurisdiction over the study area.

When humans enter the water in the vicinity of dangerous sharks, safety measures must be taken. Even experienced researchers must approach these fish with caution.

Sharks are also photographed and filmed for subsequent attentive analyses.

Underwater observations of shark predation are less common, but they do happen, for example with blue sharks preying on squid off Baja California.

In order to collect data on behaviour, interactions, feeding frequency and hunting strategies of sharks, researchers tag sharks with sonic transmitters or pop-up transmitters, and record data such as swimming speed, depth, temperature of the surrounding water, and stomach temperature. When a shark is tagged with a sonic transmitter, it sends acoustic pulses that can be detected by a hydrophone, so the shark can be tracked and monitored in real time. The pop-up tag is a satellite tag that remains attached to the shark for a long time (usually from six months to one year), and then it detaches automatically from the shark. Once popped-off, all stored data is downloaded via satellite to the researcher computers.

Sharks are tagged from a boat. A shark is lured to the surface and induced to approach close enough for researchers to tag it. The dart with the transmitter is inserted into the dorsal musculature of the shark, below the first dorsal fin or between the first and second dorsal fins. Another method of tag attachment consists of inducing the shark to swallow a fish or a piece of meat containing a hidden transmitter, which places the transmitter in the shark's stomach. White shark specialist John E McCosker suggested that a temperature elevation of the shark stomach would indicate feeding on a warm-bodied pinniped. Therefore, by inserting a temperature-sensitive transmitter into a shark stomach, it is possible to detect feeding from stomach temperature elevation. The shark stomach temperature increases to 37°C when the predator eats a pinniped, and this temperature elevation is retained for a period of time. A slow increase in shark stomach temperature may be due to ingesting a cold-bodied fish or scavenging on carrion.

Another method of observing shark predation consists of attaching a small video camera to the shark. The video recording provides a direct observation of shark predatory tactics. Useful information also comes from video recordings obtained by a small video camera attached to a favourite shark prey, such as a juvenile seal.

Great white sharks prefer coastal waters because the density of juvenile seals on which they prey is highest close to shore. At Southeast Farallon Islands and Año Nuevo Island, California, USA, great white sharks usually remain in a shallow zone that extends less than one kilometre offshore and surrounds the pinniped colonies. The small size of these areas has made it possible to conduct detailed research on great white sharks and their hunting strategies in these waters. At Año Nuevo Island, researcher A Peter Klimley and colleagues tagged some sharks with ultrasonic transmitters, and recorded location, swimming speed (burst swimming is indicative of chasing prey), depth and stomach temperature of these individuals. They simultaneously recorded shark movements and behaviour towards one another. The sharks were tracked over a 15-day period, and were monitored intensively in order to identify bouts of feeding and to describe the pattern of swimming associated with prey capture. Continuously tracking these individuals, researchers acquired a large amount of information about predatory strategies of great white sharks, and quantified the level of association between great white sharks while hunting for prey.

EXAMINATION OF SIGNS OF PREDATION AND SCAVENGING

Examination of signs of shark predation and scavenging on carcasses and wounded animals is an additional method for studying shark feeding biology. Sharks bite deeply into the soft flesh of their prey, and shark teeth usually inflict terrible lacerations on their victims. If an animal escapes and survives a shark attack, the scars can last a lifetime. Sometimes prey, dead or alive, can be found

Jaws and teeth of predatory sharks from the geographical area where the attack or the scavenging has occurred are examined, to determine the species responsible for inflicting lacerations observed on carcasses and wounded animals (photograph by Nicola Allegri).

stranded, or observed floating at the sea surface. These animals are often thought to have been hit by a boat, since the almost parallel lacerations are attributed to boat propellers.

Shark bite scars and fresh wounds are used to identify species of sharks responsible for predation and scavenging on various animals; but proper identification of the species that has inflicted a wound is not easy. Bites are measured and photographed. Jaws and teeth of predatory sharks from the geographical area where the attack or the scavenging occurred are examined, to determine the species responsible for inflicting the lacerations. The bite scars are compared with shark jaw collections held by institutes and museums. Tooth shape, size and spacing differ among species, but owing to variations in age and individual size, tooth shape, attack direction and prey movements, the characteristics of each shark bite may vary greatly. For example, the width of the bite depends on how the shark bit the prey. If the predator bit its prey using only the anterior part of the jaws, then the bite would be very small, so it is necessary to count the tooth punctures in the bite. The spacing of the cuts suggests the

Estimated 3.5m bottlenose dolphin (*Tursiops truncatus*) photographed near Lampedusa, Isole Pelagie, Italy, showing two fresh white shark bites in the dorsal region (photograph by Emiliano D'Andrea / Necton Marine Research Society).

tooth placement. An enormous 'single bite' can actually be formed by two or three overlapping smaller bites. Moreover, wounds inflicted by most shark species have not been well documented. Consequently, the species identification and the size of the responsible shark must be evaluated cautiously.

Sometimes a tooth can be lost when the shark is feeding. In rare cases tooth enamel fragments removed from the wounds of the victim are tangible evidence of the shark responsible. The fragments are usually small apical parts, left embedded in the victim's skeleton, and are often undetected. The teeth of a few sharks have sufficient characteristics to identify the attacker, which can be determined from even a single tooth or a small fragment. Sometimes great white shark bite scars and teeth fragments are also found on surfboards and boats after an attack has occurred, leaving a clear imprint of shark teeth in the fibreglass and wood. Shark bite scars found on human victims are also of importance in the identification of the species responsible for attacks.

Great white shark bite scars and fresh wounds are often found on live, moribund and dead pinnipeds such as northern elephant seals (*Mirounga angustirostris*), California sea lions (*Zalophus californianus*), Australian fur seals (*Arctocephalus pusillus doriferus*) and South African fur seals (*Arctocephalus pusillus*). Sometimes one of these pinnipeds with fresh great white shark bites is found dead on a beach. Scars are more frequently found on young individuals, and scars are generally located on the ventral part of the body. Shark bite scars are also frequently found on dolphins, such as bottlenose dolphins (*Tursiops truncatus*). Among the species responsible for these attacks are the great white shark, tiger shark, bull shark (*Carcharhinus leucas*) and dusky shark (*Carcharhinus obscurus*). Large sharks often steal parts of captured big fish such as tuna and swordfish (*Xiphias gladius*), imprinting the tangible evidence of their teeth.

Sometimes shark bite scars are also found on live tuna. These scars, usually located on the ventral part of the body, have often been inflicted by great white sharks or shortfin makos (*Isurus oxyrinchus*). Shark bite scars are also found on smaller prey, but the possibility of a conclusive identification from a scar on a relatively small-sized animal, such as a crustacean, is far less common.

Great white shark bite wounds are distinguished by large arcs measuring ten to sixty centimetres, with the characteristic tooth marks that are relatively widely spaced. In numerous cases, tooth punctures, slashes, and parallel marks of semicircular sets of cuts inflicted by partial contact with one or both jaws are present. The parallel marks are sometimes mistaken for slashes caused by boat propellers. A pattern of grooves left on bone is often indisputable evidence of the highly serrated edge of great white shark teeth. Many requiem sharks (family Carcharhinidae) have serrated teeth, but their shape and position are different. Often there is overlap between adjacent teeth, the teeth are more numerous, the upper teeth have finer serrations, and the lower teeth are narrower and without serration or with very fine serration. Furthermore, most requiem sharks have smaller bites, and the wounds are clean-cut.

A 5cm-wide hole where flesh and skin have been gouged out is tangible evidence of the cookiecutter shark (*Isistius brasiliensis*) bite. These particular rounded bites are often found on live cetaceans, elephant seals (*Mirounga* sp.), tunas, marlins and large sharks. These marks are often numerous and well-healed, and were long mistaken for damage caused by invertebrate parasites or bacteria.

When researchers find a dying animal with very recent shark wounds, they can attribute death to shark bites. But when they find a dead animal with shark bites, it is often impossible to find indications as to whether the shark bites have occurred before or after death.

A VARIED DIET

Sharks are carnivorous, or flesh-eating animals. They feed on a wide variety of prey from many different habitats, catching throughout the water column. Most sharks feed mainly on live prey, some usually attacking healthy prey (if they are able to catch them); and some typically feed on diseased or wounded animals (diseases and parasites make animals more vulnerable to shark attacks). Other sharks feed often on dead animals. All sharks usually prey on animals of comparable or smaller size. Sharks eat bony fish, cartilaginous fish, molluscs, crustaceans, pinnipeds, cetaceans, marine turtles, sea snakes, sea birds, echinoderms, worms, sea anemones, jellyfish and other organisms. The list includes even very small animals, planktonic organisms such as euphasiids and copepods. Recent research reveals a broader trophic spectrum than previously supposed for many shark species. Particular aspects of shark diets, such as shark attacks and predation on humans, cannibalism and ingestion of inedible items, will be discussed in the following sections.

The popular belief that sharks eat everything is far from the truth. Many sharks are opportunistic feeders, meaning that they are versatile and able to utilise diverse food sources depending on the availability of each food type. When their favourite prey is scarce, they will eat other species that are locally more abundant. A particularly common species, or one that is easily captured, may dominate the diet of the opportunistic sharks. According to studies by Christian Capapé on Mediterranean sharks, most opportunistic sharks are benthic species.

However, there are also numerous sharks that show dietary preferences, and some have a quite specialised diet. The crested bullhead shark (*Heterodontus galeatus*) feeds heavily on sea urchins; the common thresher (*Alopias vulpinus*) feeds mainly on anchovies, hakes and mackerels; the bonnethead (*Sphyrna tiburo*) feeds primarily on crabs and shrimps; while the great hammerhead (*Sphyrna mokarran*) favours stingrays.

Other species have a highly specialised diet. For example, the sicklefin weasel shark (*Hemigaleus microstoma*) feeds almost only on cephalopods and has a preference for octopi. Peter Klimley has suggested that the great white shark (*Carcharodon carcharias*) favours an energy-rich fat diet. In fact, these sharks show preference for elephant seals and harbour seals over sea lions and sea otters, for pinniped pups over adults, and for baleen whale blubber over muscle. Even blue sharks (*Prionace glauca*) seem to feed selectively on the energy-rich blubber layer when scavenging a whale carcass.

Australian fur seal pups (*Arctocephalus pusillus doriferus*). Many
sharks show dietary preferences: pinnipeds are a favourite food for
great white sharks (*Carcharodon carcharias*), and these predators
prefer pinniped pups to adults (photograph by Vittorio Gabriotti).

Sharks are known to live principally on fish. Bony fish are extremely common and numerous, qualities that make them highly attractive as prey to sharks. In fact, bony fish are the preferred food for most sharks, and supply a very important part of the diet for almost all sharks. Prey ranges from small pilchards to the large bluefin tuna (*Thunnus thynnus*), but the small species are the most common prey.

Examination of shark stomachs has shown that, for most species, about 70% to 80% of the diet consists of bony fish. In the north-west Atlantic Ocean, bony fish comprise 91% of the porbeagle (*Lamna nasus*) diet by mass. Off the north-east coast of the USA, bluefish (*Pomatomus saltatrix*) are the most important food of the shortfin mako, comprising 77% of its food by volume. In the Gulf of Alaska, 93% of the diet by mass of the Pacific sleeper shark (*Somniosus pacificus*) consists of bony fish, and its preferred prey is the arrowtooth flounder (*Atheresthes stomias*). Off South Africa, bony fish comprise 41% to 58% of the diet by mass of the bronze whaler shark (*Carcharhinus brachyurus*). The dominant bony fish in the diet of great white sharks and bronze whaler sharks

The bluefin trevally (*Caranx melampygus*). Bony fish supply a very important part of the diet for almost all sharks (photograph by Vittorio Gabriotti).

Sharks are an important prey item of numerous shark species (photograph by Walter Heim).

from South Africa is the small South African pilchard (*Sardinops sagax*). Off KwaZulu-Natal, South Africa, bony fish are also the most important prey group of the pigeye shark (*Carcharhinus amboinensis*). Geremy Cliff and colleagues examined a great white shark that had 477 pilchards in its stomach. The bull shark (*Carcharhinus leucas*) feeds on a wide variety of bony fish, and off KwaZulu-Natal shows preference for the Moçambique tilapia (*Oreochromis mossambicus*). These data demonstrate the importance of bony fish in the diet of sharks.

Elasmobranchs are the most important prey item of numerous shark species. Many sharks seem to prefer cartilaginous fish to bony fish. Surely sharks are the most important natural predator of other sharks (with just one exception: *Homo sapiens*). Most sharks have no enemies except other sharks and man, and sharks are also the main predators of rays. For example, sharks and rays are the most common prey of great white sharks and shortfin makos (*Isurus oxyrinchus*) in KwaZulu-Natal, South Africa. Here, the favourite white shark prey is the dusky shark (*Carcharhinus obscurus*), while the preferred mako prey is the milkshark (*Rhizoprionodon acutus*), but makos also feed heavily on dusky sharks and spotted eagle rays (*Aetobatus narinari*).

Elasmobranchs are also the most frequently encountered prey in great white sharks from California. By far the most common prey of the great hammerheads are rays. Eagle rays are also prey of other shark species such as the silvertip sharks (*Carcharhinus albimarginatus*). Off the Eastern and Western Cape of South Africa, the most important prey species of the broadnose sevengill sharks (*Notorynchus cepedianus*) are the smooth-hound (*Mustelus mustelus*), shortnose spurdog (*Squalus megalops*), leopard catshark (*Poroderma pantherinum*) and the striped catshark (*Poroderma africanum*). The dusky shark feeds often on stingrays (family Dasyatidae). Even the bull shark shows preference for elasmobranchs, especially young sandbar sharks, guitarfish (family Rhinobatidae) and stingrays. The diet of some deep-water sharks, such as the bluntnose sixgill shark (*Hexanchus griseus*), also includes chimaeras. Some sharks also feed on elasmobranch egg cases. Different sharks consume different elasmobranchs, but the percentage of this kind of prey is typically high.

Most sharks also eat a low percentage of invertebrates, such as cephalopods and crustaceans. For example, although bony fish predominates the shortfin mako diet (almost 70%), squid contributes about 15% overall. However, invertebrates, especially molluscs and crustaceans, supply a more important part of the diet for other sharks. The small-spotted catsharks (*Scyliorhinus canicula*) feed primarily on invertebrates including hermit crabs, cockles and whelks. Crustaceans are the most important prey item of the smooth-hound, which favours mantis shrimp (*Squilla mantis*), and also of the brown catshark (*Apristurus brunneus*), black dogfish (*Centroscyllium fabricii*), filetail catshark (*Parmaturus xaniurus*) and the bonnethead shark. Off the west coast of South Africa, the smalleye catshark (*Apristurus microps*) feeds heavily on euphausiid (*Euphausia lucens*). Off South Africa, molluscs are an important prey item of the bronze whaler shark (*Carcharhinus brachyurus*), comprising 30% to 35% of its food by mass.

In the same waters, dusky sharks feed heavily on chokka squids (*Loligo vulgaris reynaudii*), that comprise up to 36% of its stomach contents by mass. Cephalopods also form the main diet item of the blue shark, young bluntnose sixgill sharks and deep-water species, such as the frilled shark (*Chlamydoselachus anguineus*). In the waters of New South Wales, these invertebrates are also the most common prey of the smooth hammerhead (*Sphyrna zygaena*). A well-known octopus-eater is the whitetip reef shark (*Triaenodon obesus*). However, the diet

A jumbo squid (*Dosidicus gigas*). Cephalopods supply an
important part of the diet for many sharks (photograph
by Phil Zerofski).

of some sharks is made up almost entirely of invertebrates. For example, the
leopard shark (*Triakis semifasciata*) feeds heavily on innkeeper worms (*Urechis
caupo*), crabs and clams, while the sicklefin weasel shark eats cephalopods,
crustaceans and echinoderms.

Some sharks show a marked affinity for marine mammal meat. We know that
the following sharks feed on cetaceans: great white shark, shortfin mako, pigeye
shark, bronze whaler shark, Galapagos shark (*Carcharhinus galapagensis*),
bull shark, blacktip shark (*Carcharhinus limbatus*), oceanic whitetip
shark (*Carcharhinus longimanus*), dusky shark, sandbar shark, tiger shark
(*Galeocerdo cuvier*), blue shark, hammerhead shark (*Sphyrna* sp.), bluntnose
sixgill shark, broadnose sevengill shark, Portuguese shark (*Centroscymnus
coelolepis*), Greenland shark (*Somniosus microcephalus*), Pacific sleeper shark,
and cookiecutter shark (*Isistius brasiliensis*). This is quite a long list, but while
some of these sharks feed on both live and dead marine mammals, others may
feed on dead individuals only.

In fact, most sharks eat these animals only when they find a carcass. However,
some species, such as the great white shark, tiger shark and cookiecutter shark,
feed often on live marine mammals. For example, the tiger shark is believed to
be the primary predator of Hawaiian monk seals (*Monachus schauinslandi*).

A bottlenose dolphin (*Tursiops truncatus*). Many sharks feed only on dead cetaceans, but a few, such as the great white shark (*Carcharodon carcharias*) and tiger shark (*Galeocerdo cuvier*), attack live individuals (photograph by Vittorio Gabriotti).

Consequently tooth shape and predatory behaviour of these shark species is specialised for eating marine mammals. By contrast, the shortfin mako rarely feeds on cetaceans.

Predation on adult mysticetes and very large odontocetes has not been documented. As with many species, the number of potential predators shrinks as animals increase in size. There are no observations of sharks preying on very large cetaceans, with the exception of the cookiecutter shark (see 'Ambush hunters', page 124). However, these immense animals are commonly eaten by sharks after their death.

The sharks identified as able to attack and kill small odontocetes are the great white shark, shortfin mako, tiger shark, bull shark, oceanic whitetip shark and dusky shark. Sharks must be considered significant predators and scavengers of small cetaceans in many areas. Great white shark predation on dolphins is more frequent in areas where pinnipeds are less abundant or absent, such as the Mediterranean Sea, since this shark is an opportunistic predator. Shark attack scars were found on 36% of 334 individually identified bottlenose dolphins (*Tursiops truncatus*) in Moreton Bay, Queensland, Australia; on 22% of specimens examined in Sarasota Bay, Florida, USA; and on at least 10% of 145 specimens caught off KwaZulu-Natal, South Africa. Based on stomach contents four sharks were identified as the most common predators on bottlenose dolphins off the KwaZulu-Natal coast; these were the bull shark, dusky shark, tiger shark and great white shark.

Some sharks also eat a low percentage of marine reptiles and sea birds. Sea turtles are an occasional prey item of the tiger shark, great white shark and bull shark. The tiger shark is also one of the most important predators of sea snakes. Birds are an occasional food item of the tiger shark, great white shark, shortfin mako, bull shark and the Galapagos shark.

Unlike bony fish that drink sea water, sharks do not. The salt content of shark tissue is higher than that of bony fish, as well as of sea water, owing to the shark's retention of high concentrations of certain metabolic wastes, including urea, in their bodies. Therefore, water naturally diffuses into the body through external cell membranes by osmosis (the passage of water across a semipermeable membrane from a dilute solution to a more concentrated solution), and they do not need to drink sea water.

A grey heron (*Ardea cinerea*) watches a blacktip reef shark (*Carcharhinus melanopterus*) venturing into shallow waters of Felidhu atoll, Maldive Islands. Birds are an occasional food item of some sharks (photograph by Vittorio Gabriotti).

CHANGES IN DIET

Considerable variation exists in the diet of a given species of shark. In fact, a shark's diet is closely related to its size and age, as well as to geographic location and the season.

Diet as related to size means that, in general, larger sharks can eat the largest prey. Consequently, larger prey become increasingly important in the diet of sharks as they grow larger. Therefore there is great variation in the diet of many sharks as they grow.

As stated previously, variation in the diet is accompanied by a variation in tooth shape, therefore these are ontogenetic dietary changes. Since young sharks would have a hard time capturing, cutting and ingesting large prey, these changes are a function of prey handling capability. Bony fish, being usually smaller than cartilaginous fish, and some other prey items such as molluscs are more easily captured and ingested by smaller sharks. Therefore many young sharks feed mainly on bony fish and molluscs. But the incidence of this kind of prey declines as the sharks attain larger sizes and target larger prey, such as cartilaginous fish and marine mammals, whose incidence thus increases.

Analysis of prey type in relation to great white shark size shows that small individuals measuring less than 3m in length feed primarily on fish (especially bottom-dwelling forms), while larger sharks have been identified as important predators and scavengers on marine mammals (pinnipeds, dolphins and whale carcasses). Stomach contents from 54 young great white sharks caught in the New York Bight showed a high incidence of demersal fish, especially sea robins, as well as pelagic fish, such as menhaden (*Brevoortia* sp.), and also hake (*Merluccius* sp.), bluefish (*Pomatomus saltatrix*), flounders, skates, smoothhound sharks (*Mustelus* sp.), crabs and starfish.

Young white sharks have teeth that are more narrow, finely serrated or almost smooth, resembling those of the porbeagle (*Lamna nasus*). These teeth are better adapted for grasping than for cutting, hence the young individuals feed on smaller prey which they usually swallow whole. Larger sharks have teeth that are broader and heavily serrated, adapted for cutting prey into smaller pieces or for neatly excising a mouthful of flesh from large prey, which they often eat in pieces. The tiger shark (*Galeocerdo cuvier*) also shows changes in diet with size. Larger tiger sharks are equipped with jaws of enormous power and target sea turtles, while smaller sharks eat sea snakes.

The major prey of the shortfin mako is fish. Accounts of shortfin makos killing and eating large swordfish (*Xiphias gladius*) of 180 kilograms or more

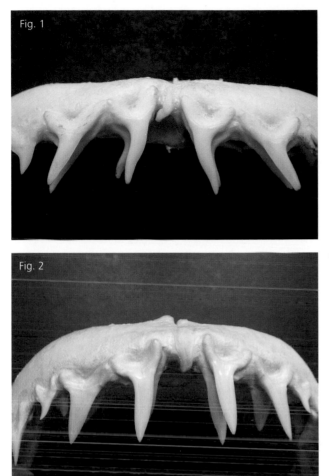

Fig. 1

Fig. 2

Tooth shape alters as sharks grow larger and feed on different animals. The shortfin mako (*Isurus oxyrinchus*) is born with narrow teeth (Fig. 1) but as it grows these become thicker (Fig. 2) (photograph by Alessandro De Maddalena).

are common, but the shift to this large prey occurs when shortfin makos attain weights greater than 130 kilograms. Adult smooth hammerheads (*Sphyrna zygaena*) feed on sharks, rays and stingrays, while a primary food source of the young individuals consists of cephalopods. Analysis of the prey of bull sharks (*Carcharhinus leucas*) shows that individuals measuring less than 1.4m in length feed mainly on bony fish, but their incidence declines when the sharks attain larger sizes and target cartilaginous fish, marine mammals, marine turtles and birds.

Dusky sharks (*Carcharhinus obscurus*) and bronze whaler sharks (*Carcharhinus brachyurus*) measuring over 2m feed on elasmobranchs more often than do young individuals. Stomach contents from adult bronze whaler sharks caught off the Eastern Cape coast of South Africa show an important preference for octopi, while these molluscs are absent from young shark stomachs. This difference suggests a variation in the bronze whaler shark diet, with an increase in benthic feeding as they grow larger.

In Elkhorn Slough, California, leopard sharks (*Triakis semifasciata*) measuring less than 70cm in length feed mainly on fish, but the incidence of fish declines when the sharks attain larger sizes and eat bony fish and their eggs, along with clams and innkeeper worms (*Urechis caupo*).

Young bluntnose sixgill sharks (*Hexanchus griseus*) measuring less than 1.2m in length feed almost entirely on cephalopods and bony fish, while individuals between 1.2 and 2m in length mainly prey on bony fish, cartilaginous fish and cephalopods. Finally, individuals over 2m feed primarily on cetaceans and bony fish.

Sharks frequently give birth to their young in nursery areas, often coastal waters, lagoons or estuaries, where juveniles remain for the early period of their lives. Newborns immediately learn to hunt for food. In the nursery areas, the juveniles encounter few predators. The abundance of suitable prey is the major reason some regions function as nursery areas for sharks. In this restricted habitat, juveniles may encounter a low diversity of prey. Therefore, young sharks may have a restricted diet, and an abundant species can become their most common prey item. David A Ebert reported that young bluntnose sixgill sharks primarily consume cephalopods that are particularly abundant in their nursery area off southern Namibia. Once bluntnose sixgill sharks grow older, molluscs become of secondary importance in their diet, and as they begin to mature the sharks move offshore and discover other food sources. Therefore, the diversity of feeding habits in larger sharks may reflect the broader range of habitats in which they live.

There is considerable variation in the diet of numerous sharks from one location or season to the next. Regional differences in the diet are attributable to the higher incidence of particular prey. In fact, many sharks focus their hunting activity on the most abundant local species. In many areas, the diet of shortfin makos (*Isurus oxyrinchus*) is made up mainly of bony fish, and the incidence of elasmobranchs in its diet is very low. But in offshore waters where depths are greater than 180m, these sharks eat more cephalopods, and in KwaZulu-Natal, South Africa, their most common prey is small sharks and rays. In the north-western Atlantic Ocean shortfin makos feed heavily on bluefish (*Pomatomus saltatrix*), while this species is absent from shortfin mako stomachs analysed in KwaZulu-Natal, South Africa, and New South Wales, Australia. This is strange, because bluefish are present in all these waters. Geremy Cliff and colleagues have suggested that this difference is attributable to the fact that KwaZulu-Natal bluefish are smaller than north-western Atlantic bluefish on which makos feed.

Along the USA Pacific coast, northern anchovy (*Engraulis mordax*) is the most important species in the diet of the common thresher shark (*Alopias vulpinus*) south of 34°N latitude, while north of this latitude the most important prey is Pacific hake (*Merluccius productus*). Off New South Wales, Australia, cephalopods are the most important food of the smooth hammerhead (*Sphyrna zygaena*), while in South Africa the primary food source of this species consists of bony fish.

At the Farallon Islands and Año Nuevo Island, California, USA, great white sharks (*Carcharodon carcharias*) kill more seals during the fall and winter, because at this time young elephant seals of one to two years of age (the preferred prey for the great white shark) are most common in these waters. In many locations of the world, such as California, South Africa and South Australia, elasmobranchs and pinnipeds are the major components in the diet of great white sharks, while in the Mediterranean Sea, where there are virtually no pinnipeds and sharks are not abundant, great white sharks feed mainly on cetaceans, bony fish (particularly tuna and swordfish) and marine turtles.

Where there are few or no preferred prey, the sharks find something else to eat.

Off southern Africa, David A Ebert observed that the bluntnose sixgill shark feeds primarily on cephalopods, bony fish, cetaceans and chondrichthyans (with a marked incidence of small sharks and chimaeras). The author, together with Antonio Celona and Teresa Romeo, examined the diet of the bluntnose sixgill shark in eastern north Sicilian waters, Italy. We discovered that in this area the large predator feeds primarily on bony fish, while cetaceans, sharks and chimaeras were completely absent from our sample. Strangely, this regional difference in the diet of the bluntnose sixgill shark does not seem to be attributable only to a different incidence of the prey, as both dolphins and small sharks are very common in the study area.

In Catalonian waters off Spain, kitefin sharks (*Dalatias licha*) eat more small sharks during the spring and winter, more crustaceans during the summer, and more cephalopods during the autumn. Certain sharks feed on schooling migratory fish when they migrate across their area, and when these prey move to other waters the predators must feed on other animals. In fact, many species eat animals seasonally most abundant, and a particularly common prey may dominate their diet. In the western North Atlantic, shortfin makos consume bluefish (*Pomatomus saltatrix*) during winter and spring, and consume more cephalopods during summer since bluefish move inshore at that time of the year. A varied diet is advantageous to sharks because population density, habitat, geographical distribution and seasonal movements of a given shark species are not strongly limited by food availability.

CANNIBALISM

Many sharks eat members of their own species; several species have been observed to attack, kill and eat smaller or wounded members of the same species. For example, the tiger shark (*Galeocerdo cuvier*), Galapagos shark (*Carcharhinus galapagensis*) and bull shark (*Carcharhinus leucas*) are well-known cannibalistic sharks. Other sharks that occasionally eat members of their own species include the blue shark (*Prionace glauca*), dusky shark (*Carcharhinus obscurus*) and the great white shark (*Carcharodon carcharias*). Cannibalism is particularly common among requiem sharks (family Carcharhinidae). Interactions that are non-fatal are manifested by lacerations such as parabolic puncture marks caused by the bite of another shark.

A proportion of shark attacks on other sharks are non-predatory in motivation, principally being the result of reproductive or antagonistic behaviour. Sharks, especially whaler sharks (genus *Carcharhinus*), have often been observed to attack conspecifics when stimulated by blood and food in the water.

Inexperienced juvenile sharks are especially vulnerable to predation, and may easily fall prey to larger members of their own species. Aquarists have observed captive horn sharks (*Heterodontus francisci*) lay horny egg cases on the bottom

Bull sharks (*Carcharhinus leucas*) occasionally eat members of their own species (photograph by Harald Baensch).

Female blue sharks (*Prionace glauca*) stop feeding temporarily when they have to give birth to pups, while mature males move far out of the area (photograph by Walter Heim).

of the tank, then suck out and eat the content. Hence, female sharks have their pups in nursery areas to protect their young from being eaten by male sharks. As stated previously, areas where only newborns live have been observed for numerous shark species.

According to some researchers, females stop feeding temporarily when they give birth to their pups, while mature males move far out of this area. Researchers have hypothesised that this behaviour may exist in numerous species, including the bluntnose sixgill shark (*Hexanchus griseus*), broadnose sevengill shark (*Notorynchus cepedianus*), bigeye thresher shark (*Alopias superciliosus*), shortfin mako (*Isurus oxyrinchus*), longfin mako (*Isurus paucus*), porbeagle (*Lamna nasus*), tope shark (*Galeorhinus galeus*), bronze whaler shark (*Carcharhinus brachyurus*), blue shark (*Prionace glauca*) and smooth hammerhead (*Sphyrna zygaena*).

As juvenile sharks mature, they move far away from the nursery area. When fully mature, many large sharks enjoy partial safety from predation (sharks usually attack animals of smaller size). Moreover, males of some species stop feeding during the mating season. These behaviours reduce the risk of cannibalism, and for similar reasons adult sharks often segregate by size.

Writer and shark fisherman, William Travis, observed that sharks left the zone when dead sharks or shark heads were dumped overboard. Species that showed this behaviour were sandtiger sharks (*Carcharias taurus*), hammerhead sharks (*Sphyrna* spp.) and whaler sharks (genus *Carcharhinus*). Travis reported that sharks reacted in this way only in the presence of heads and complete specimens, while beheaded specimens, fins, livers and other organs were ignored or attracted more sharks. Travis hypothesised that shark heads contain some chemicals that act as a shark repellant. Shark specialist Stewart Springer observed that decomposing sharks on a line can keep other sharks away.

Cannibalism has also been observed in shark embryos (intra-uterine cannibalism). At least an aplacental viviparous species, the sandtiger shark (*Carcharias taurus*), is embryophagous: the first embryo to hatch within the uterus kills and eats its siblings (see 'Reproduction and life span' on page 18).

The tiger shark (*Galeocerdo cuvier*) is probably the more omnivorous shark, so it is not surprising that it is a well-known cannibalistic species (photograph by Wolfgang Leander).

SHARK ATTACKS ON HUMANS

Sometimes sharks attack even human beings; people are occasionally killed or injured by these animals. These cases provoke extreme emotional reactions from most people, and the media irrationally amplify each incident.

As a result of several programmes of data collection, substantial information about historical and recent shark attacks has been collected. These programmes include specific research on shark attacks, such as the Global Shark Attack File (GSAF, a compilation of shark attacks worldwide based at the Shark Research Institute in Princeton, New Jersey, USA), the International Shark Attack File (ISAF, a compilation of shark attacks worldwide based at the Florida Museum of

The great white shark (*Carcharodon carcharias*) is the sea's most feared predator. This species is responsible for the highest number of attacks on humans (photograph by Vittorio Gabriotti).

Natural History in Gainesville, Florida, USA) and the Australian Shark Attack File (ASAF, a compilation of shark attacks in Australian waters based at the Taronga Zoo in Sydney, Australia).

Other data have been collected by scientists in more general research programmes that include an important part on shark incidents: for example, the Shark Research Committee (SRC, a research organisation studying all shark-human interactions along the Pacific coast of North America with particular emphasis on shark attacks in this zone, based in Van Nuys, California) or the Italian Great White Shark Data Bank (a programme of data collection on the great white shark in the Mediterranean Sea, including a compilation of great white shark incidents in these waters, based in Milan, Italy). The data of these research programmes are continually updated.

Despite widespread publicity by the media, compared to the total number of deaths from any other form of water-related activity, deaths caused by sharks are very low. The fearsome reputation of these animals is exaggerated. Despite the large number of people who frequent beaches and seas, and the fact that dangerous sharks are present in all waters, shark attacks are extremely rare in most parts of the world. Human beings are not a usual part of any shark diet, and the human attack rate is very low. In most cases, the attack ends after the initial contact, and the shark does not eat or kill the victim. Most attacks result in significant blood loss, not massive consumption by the shark. We presume that sharks do not regard humans as food, and that most attacks are not motivated by hunger.

The human-shark attack fatality rate is low, and when the victim dies it is usually as a result of shock, blood loss or other injury. According to the International Shark Attack File, worldwide there are 70 to 100 shark attacks annually and only five to 15 are fatal.

It is surprising that so few incidents occur. This is a very low rate taking into account the wide distribution of most dangerous sharks and the high number of people who swim and dive in the seas of the world. However, the actual number of shark attacks around the world per year is unknown, and the incidents recorded represent only a portion of the actual total. In fact, in many areas several shark attacks go unrecorded.

Numbers of shark attacks are rising each year because of the increasing worldwide population of humans, and the consequently increasing numbers of bathers, surfers, divers, anglers, fishermen and boats in the water.

Shark attacks occur in shallow, deep, warm and cold waters, but more attacks occur where weather is favourable for recreational swimming. The majority of incidents occur in Florida, USA, partly because of the popularity of these waters for swimming and surfing, and partly because many potentially

Lemon sharks (*Negaprion brevirostris*). Shark attacks occur in all kinds of water: shallow, deep, warm and cold (photograph by Harald Baensch).

dangerous sharks inhabit this area. However, in many Third World countries, shark attacks are rarely reported. Moreover, the author is aware of the fact that in some countries reports of attacks are intentionally suppressed in order to avoid extreme reactions from the media and damage to the tourism industry.

Most attacks occur in water depths of between 1.5 and 3 metres. Large dangerous sharks occasionally come close inshore near populous locations, and are encountered more often in zones where the bottom drops off very rapidly. Islands, straits, channels and shoals are also likely attack sites. The large predators swim in these areas because their prey also congregate there.

Many sharks are large and powerful enough to inflict serious wounds or to kill a human, but most species are actually inoffensive to humans, and only a few show aggressive behaviour and are a potential threat to human beings. Only three species are involved in most incidents: the great white shark (*Carcharodon carcharias*), tiger shark (*Galeocerdo cuvier*) and bull shark (*Carcharhinus leucas*).

The great white shark is responsible for the highest number of attacks. The bull shark is particularly dangerous for its habit of entering freshwaters. These three sharks are powerful, very large and massive; they have large teeth with serrated margins, and a wide mouth; they feed on large prey, are the more omnivorous species, occur in a great variety of habitats, and are widely distributed in the seas throughout the world.

Sometimes these species are extremely aggressive to humans without provocation, and therefore constitute a great danger if encountered by a swimmer or a diver. In some areas these sharks are quite common and they swim relatively close to shore (particularly the bull and tiger sharks), where they could easily catch and eat many bathers. But even for these three species, human beings are a rare and accidental prey. Moreover, many cases have been reported of great white sharks, tiger sharks and bull sharks approaching divers and bathers closely without showing any aggressive behaviour.

Other species that are known to become aggressive towards humans are the blue shark (*Prionace glauca*), shortfin mako (*Isurus oxyrinchus*), great

Caribbean reef sharks (*Carcharhinus perezi*) are known to become aggressive towards humans (photograph by Harald Baensch).

Most potentially dangerous sharks are timid when interacting
with humans under normal conditions when food is not present,
but should be treated with caution (photograph by Walter Heim).

hammerhead shark (*Sphyrna mokarran*), oceanic whitetip shark (*Carcharhinus longimanus*), lemon shark (*Negaprion brevirostris*), dusky shark (*Carcharhinus obscurus*), grey reef shark (*Carcharhinus amblyrhynchos*), bronze whaler shark (*Carcharhinus brachyurus*), blacktip reef shark (*Carcharhinus melanopterus*), blacktip shark (*Carcharhinus limbatus*), Galapagos shark (*Carcharhinus galapagensis*) and the Caribbean reef shark (*Carcharhinus perezi*). This list is incomplete, since there are other species, particularly other whaler sharks (genus *Carcharhinus*), that may be involved in attacks on humans.

Only a portion of the attacks can be attributed to a particular species, because post-attack identification of the offending shark is difficult. In most cases, the definitive identification of the species responsible for an attack can only be made if tooth fragments are found in the victim's wounds, if the specimen is photographed or captured, or if a witness is a marine life expert. Very few people know one species from another, and the description given by the victim or witnesses is often insufficient or unreliable; commonly the animal is simply described as a 'large grey shark'. Consequently, in most cases, the species involved is never identified. This problem is more evident in the case of whaler sharks, since these species are very similar.

The nature of shark incidents, and the circumstances surrounding each case, vary widely. However, all shark attacks can be divided into two categories; provoked and unprovoked attacks.

When a person's behaviour elicits aggression, the attack is considered provoked. Most sharks are quiet animals that ask only to be left alone. However, we have to remember that almost every shark species is capable of inflicting painful wounds when slightly molested or injured. For example, there have been cases in which divers have drowned because a quiet nurse shark (*Ginglymostoma cirratum*) has bitten them and refused to open its powerful jaws. Most provoked attacks involve scuba divers or sport and commercial fishermen catching sharks.

Most potentially dangerous sharks are usually timid when interacting with humans under normal conditions when food is not present; nevertheless, they should be treated with caution. Especially great white sharks and shortfin mako sharks should never be harassed, as these species have been responsible for several attacks on boats, both provoked and unprovoked.

Almost all unprovoked attacks can be attributed to various causes, mainly feeding, defence and social behaviours. Many attacks have been linked with the presence of large numbers of fish or other shark prey, such as pinnipeds close to the victims. Evidence suggests that sometimes sharks may attack in order to defend themselves or their pups from possible predation. In fact, female grey reef sharks show elevated aggression in pupping areas.

Surfers are always at particular risk from shark attacks because of their habit of going far offshore. Divers with speared fish or people who are bleeding are the most likely to attract a shark. Dangerous sharks have sometimes approached spearfishing divers and stolen the captured fish without attacking the man, but in other cases they have also inflicted severe or fatal injuries on divers. When there is a lot of food in the water and many sharks are attracted, they can get very aggressive, making the situation extremely dangerous and hard to control.

Many divers want to swim unprotected and handfeed potentially dangerous sharks in their habitat. In areas such as the Red Sea, Maldives, Bahamas,

Coral Sea and French Polynesia, many divers pay to see requiem sharks (family Carcharhinidae) and other species being fed by the dive master. Recently there has been much controversy on this subject. Some shark experts believe that shark-feeding dives are an acceptable way of educating and helping people to understand these feared predators, while others think this is simply a way to increase the number of shark incidents. If due caution is not exercised, accidents can result. However, shark feeding encourages shark protection and natural environmental conservation, because for many local communities it makes these fish and their habitat more valuable intact than damaged or destroyed.

A completely different category of shark incidents involves those that occur after sea and air disasters. In the case of a shipwreck or an airplane crash in the ocean, there can be many injured people in offshore waters whose presence can attract numerous dangerous sharks. This situation becomes very dangerous and hard to control. In this type of attack, the species involved are oceanic pelagic sharks, such as the oceanic whitetip shark and shortfin mako.

It is not easy to detect common features in shark attacks. According to the International Shark Attack File, unprovoked shark attacks generally fall into three categories: 'hit and run', 'bump and bite' and 'sneak' attacks. Scavenging on corpses is not considered an 'attack'; therefore it will be discussed under 'Scavengers', page 150.

The most frequent are 'hit and run' attacks. The shark takes a bite out of a human, for example a surfer or a bather, and then swims away, inflicting much less damage than it might do otherwise. In these attacks, victims usually receive relatively small injuries, with little loss of flesh. These injuries are inconsistent with the idea that hunger is the motivation for the attack. Some researchers think that in cases like this, under conditions of poor water visibility, sharks mistake the human being for another usual prey (surfboards and wetsuited divers have often attracted the attention of sharks, perhaps because of their resemblance to large fish or pinnipeds). These researchers suggest that, upon mistaking the victim for a more acceptable food item and attacking, the predator immediately understands that it has made a mistake, and releases the person. Many of these incidents could actually be motivated by a desire to defend individual territory. The victim seldom sees the attacking shark, but it is suspected that many species may be responsible, mainly requiem sharks such as the blacktip shark, spinner shark and blacknose shark (*Carcharhinus acronotus*).

Two other kinds of shark attack are less common. These are 'bump and bite' attacks and 'sneak' attacks. Injuries to these attacked victims (usually bathers or divers) are usually very serious or fatal. In 'bump and bite' attacks, the shark usually circles and bumps the person before executing the attack. 'Sneak' attacks are characterised by the shark attacking without warning. In both types,

The great white shark (*Carcharodon carcharias*) often attacks without warning. Most of its victims do not see the predator before the attack (photograph by Vittorio Gabriotti)

the shark often bites repeatedly and deeply. The wounds are usually severe, involving removal of a considerable amount of tissue from the victim. These behaviours can be motivated by hunger. The great white shark, tiger shark and bull shark (the species most often implicated in human attacks) are responsible for many 'bump and bite' and 'sneak' attacks. Other sharks, including the oceanic whitetip shark, shortfin mako, great hammerhead, blue shark and other whaler sharks such as the Galapagos shark and Caribbean reef shark, have also been implicated in 'bump and bite' and 'sneak' attacks.

Great white sharks often bite animals that are not consumed, on many occasions killing them in the process. This happens with humans as well as with sea otters (*Enhydra lutris*), northern fur seals (*Callorhinus orsinus*) and African penguins (*Spheniscus demersus*). This behaviour is common. Attacks on these prey species by great white sharks have resulted in severe wounds and death for a substantial number of individuals. For example, long-term studies conducted by Jack A Ames and colleagues in California, USA, have shown that sea otter mortality caused by great white sharks reaches 20% per year in the Año Nuevo Island area. Timothy C Tricas and John E McCosker have suggested that great white sharks mistake surfboard silhouettes for pinnipeds. George H Burgess and Matthew Callahan reported that a high percentage of victims wear black gear or clothing similar to the dark colouration of many marine mammals. Wesley Rocky Strong Jr and colleagues have observed that these sharks prefer a seal-shaped target when presented simultaneously with a square target. However, Ralph Collier and colleagues demonstrated that great white sharks also attack inanimate objects of a variety of shapes, sizes and colours, none resembling the shape, size or colour of a pinniped. Consequently, they have suggested that these predators are in most such cases determining the suitability of the prey as food. However, American researcher Scott W Michael witnessed a shortfin mako of approximately 1.5m attacking, killing and rejecting a blue shark of approximately 80cm. Canadian shark expert R Aidan Martin hypothesised that some shark attacks may be play activities. We must conclude that the real reasons behind these attack behaviours remain largely unknown.

Meshing is the most effective method for protecting beaches from dangerous sharks. Nets are placed parallel to the shore, and sharks are captured as they try to pass through them. Shark nets protect many beaches of Australia and South Africa. However, this method is difficult and extremely expensive to establish and maintain, since nets must be patrolled, cleaned and repaired often (in South Africa this work is done by a specific organisation entitled Natal Sharks Board). Consequently, only relatively few restricted areas can be protected. Moreover, there is concern about the effects of meshing on the marine ecosystem. Many alternatives to meshing have been tested, including chemical and electric repellants, which provided good results. Other efforts include protective clothing, such as the sharkproof suit of chain mail called Neptunic.

Much about shark behaviour is still unknown, but can be at least partially understood and predicted. We know that some shark behaviour involves a communication function, and these fish often attempt to communicate with humans using particular signals before executing an attack (see also 'Competition', page 166). These particular behaviours may function as a means of defending the shark, its pups, its hunting ground, its food or its individual territory. A correct interpretation of shark behaviour, particularly the threat displays and predatory tactics, can significantly minimise the risk of an attack and prevent many incidents. More detailed research on shark behaviour should be continued.

Often the shark circles the prey before attacking. If rapidly or repeatedly approached, cornered or disturbed, grey reef sharks change their swimming style and display a well-known ritualised threat behaviour in which they swim with an exaggerated motion that often precedes an attack. The shark swims with back arched, snout lifted, mouth slightly open, and pectoral fins depressed, and sometimes turns in horizontal spirals or in figure-eights as it swims closer to the diver. The shark often interrupts the attack within a short distance from the diver and continues this behaviour until the diver leaves. The degree of display depends on the anxiety of the shark.

Other shark species, such as the blacknose shark (*Carcharhinus acronotus*), Galapagos shark and the bonnethead shark (*Sphyrna tiburo*), perform a similar threat display. Threat displays performed by the shortfin mako include gaping its lower jaw slightly and turning in figure-eights as it swims closer to the diver. Sometimes even the great white shark shows a threat display with jaws slightly open and pectoral fins depressed. Unfortunately the attacking great white shark most often does not perform any warning behaviour and is not seen before the attack; therefore incident prevention through interpretation of white shark behaviour is almost impossible.

The more knowledge obtained about the communication signals these fish use, the closer we come to resolving the problem of shark attacks. However, we cannot forget that shark attack is a rare consequence of entering the sea, and is a risk that each diver, surfer or bather must consider. This is the realm of the shark, and we use it at our own risk.

When persistently approached the shortfin mako (*Isurus oxyrinchus*) performs a ritualised threat display that includes gaping its lower jaw slightly and turning in figure-eights as it swims closer to the diver (photograph by Walter Heim).

INEDIBLE ITEMS AND VARIOUS ODDITIES

As stated previously, some sharks decide on food palatability while it is lodged in their mouths. Sharks are highly curious animals. They investigate and sometimes strike, bite and swallow many foreign objects, including unusual food and inedible items. Usually items found to be distasteful are spat out. Numerous attacks on inedible items have been observed during several studies of shark behaviour.

Inferences drawn from baited conditions could be invalid, but it is well known that, even under non-baited circumstances, some sharks often attack and swallow inedible items. Sharks can also ingest hooks when they swallow hooked fish. The author has recently observed a 10cm-long hook found by Italian ichthyologist Luigi Piscitelli in the stomach of a 1.2m young shortfin mako (*Isurus oxyrinchus*). Remains of frigate tuna (*Auxis thazard thazard*) were also found in the same specimen. The hook was probably swallowed when the mako fed on a hooked frigate tuna. Geremy Cliff and Sheldon FJ Dudley examined 19 bull sharks (*Carcharhinus leucas*) with hooks in their stomachs.

Small quantities of seaweed are found in the stomachs of some sharks, such as the blind shark (*Brachaelurus waddi*), great white shark (*Carcharodon carcharias*), bull shark, bronze whaler shark (*Carcharhinus brachyurus*) and pigeye shark (*Carcharhinus amboinensis*). Angelsharks (family Squatinidae), dusky smooth-hound (*Mustelus canis*), piked dogfish (*Squalus acanthias*), broadnose sevengill shark (*Notorynchus cepedianus*), nurse shark (*Ginglymostoma cirratum*), narrowtail catshark (*Schroederichthys maculatus*), blue shark (*Prionace glauca*), bull shark, dusky shark (*Carcharhinus obscurus*) and blind shark are also reported to feed on algae. However, algae, plants, seaweed, mud, stones and pebbles are considered unusual items in the shark diet. These materials found in the stomachs of sharks are thought to have been ingested along with food taken off the bottom of the ocean.

Malcolm J Smale and David A Ebert have observed that the occurrence of algae and seaweed in shark stomachs is usually associated with benthic animal remains, such as octopus and catshark egg cases. Sometimes the prey grasps the substratum when trying to escape, and the predator ingests both prey and algae. Strangely, despite the fact that great hammerheads (*Sphyrna mokarran*) often feed very close to the bottom, seaweed was not found in any of the stomachs of 209 great hammerheads caught off KwaZulu-Natal, South Africa.

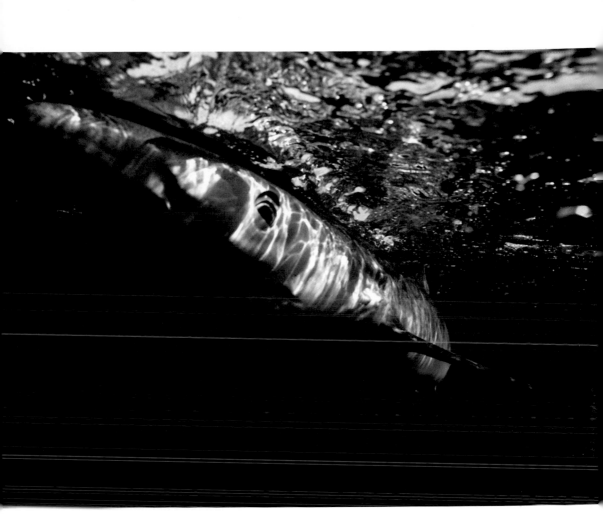

Sharks can ingest hooks when they swallow
hooked fish (photograph by Walter Heim).

Some researchers have hypothesised that stones and pebbles may be useful as
ballast. Some of these materials are also thought to have been ingested as part of
the stomach contents of the prey.

The stomachs of some sharks contain inedible items and other oddities. These
materials are rarely found in the stomachs of most species, but a few sharks,
mainly the great white shark, tiger shark and bull shark, swallow inedible items
more frequently than others. The indiscriminate feeding habits of these sharks
have been well documented. These animals have a strong tendency to strike
non-prey objects. These species are probably the more omnivorous sharks, and
in fact they eat almost anything.

The list of oddities found in the stomachs of great white sharks from various
locations (including the Mediterranean Sea, South Africa, Australia and other

This estimated 5.94m great white shark (*Carcharodon carcharias*) was caught in May 1974 off Isola La Formica, Italy. Its stomach contained a goat, plastic bottles and plastic bags (photograph: archives of the Italian Great White Shark Data Bank).

locations) includes a 7kg stone, a 2.74m wire, thirty-one 15cm hooks, pants, boots, shoes, baskets, a small board of cork, a raincoat, two or three coats, clothes, an automobile licence plate, plastic bags, garbage, a sheet of cardboard, a ship scraper, a wicker-covered scent bottle, and two pumpkins. The list also includes terrestrial animals such as a horse, sheep, calf, lamb, pig, cat, dog, goat, a whole skin of a buffalo, and men wearing suits of armour. One stomach of a great white shark from South Africa contained 24 small well-worn pebbles that weighed 50 grams; possibly this material originated from the stomach of a South African fur seal (*Arctocephalus pusillus*) eaten by the shark.

In tiger shark stomachs from various locations (including South Africa, Australia and other areas) were found coal, shoes, car licence plates, paint cans, a roll of tar paper, a hat, two 900g tins of green peas, a cigarette tin, an unopened can of salmon, a 900g coil of copper wire, plastic bags, small barrels, nuts, bolts, driftwood, a tom-tom, a wallet, boat cushions, leather, fabrics, the head and forequarters of a crocodile, hyenas, monkeys, chickens, pigs, cattle, donkeys, the hind leg of a sheep, a dog and rats. In bull shark stomachs, from South Africa and other locations, were found butcher bones, a rabbit, a mole, an antelope, cattle, sloths, rats, the head of a dog, parts of a cat and a roasted potato.

Other sharks rarely ingest unusual material, and are far less prone to swallow inedible garbage than are the great white shark, tiger shark and bull shark. In angelshark stomachs (family Squatinidae) from various locations were found a hat, a 900g can of mustard and a 45cm piece of wood bristling with nails. In an oceanic whitetip shark stomach (*Carcharhinus longimanus*) caught near Rarotonga, Cook Islands, were found three chicken heads, two steaks, a 38cm leaf and a 48cm pine needle. In the stomach of a shortfin mako caught in the Messina Strait, Italy, were found a small bow net and some large buttons sewed on pieces of a dark cloth (from the kind of coat commonly worn by seamen). In the stomachs of dusky smooth-hounds caught in Cuban waters were found chicken-heads. Greenland sharks (*Somniosus microcephalus*) have consumed entire reindeer without horns, and parts of horses. In Port Jackson shark stomachs (*Heterodontus portusjacksoni*) were found potato and orange peels. Italian marine biologist, Francesco Costa, reported that garbage has also been found in the stomachs of smooth hammerheads (*Sphyrna zygaena*). In other shark stomachs of unknown species were found 25 quart bottles of Vichy water bound together with a wire hoop, three bottles of beer, a powder puff and a wristwatch.

Most of the land animals found in the stomachs of sharks were not actually hunted by the sharks, but were discarded into the sea by humans, or drowned (see 'Scavengers', page 150). Nevertheless, we know that sharks can occasionally attack live terrestrial mammals such as horses, dogs and hippos.

Some researchers believe that sharks mistake inedible items for some usual prey but, as stated previously, Ralph Collier and colleagues demonstrated that great white sharks attack numerous inanimate objects, none resembling a prey animal. The objects observed included conical boats, circular crab trap buoys, rectangular float bags and an inflatable boat. These studies indicate that great white sharks approach and seize inedible items without regard to shape, colour and size. Sharks may simply swallow these oddities when they are determining potential food value. Even if vision is important in discriminating sources of food, it seems that great white sharks decide on food palatability while it is lodged in the mouth. After the initial attack, the inedible item is usually rejected, but sometimes the predator may ingest these oddities in error. Sometimes it may also eat inedible objects accidentally along with food. While sharks are also attracted to metal items owing to electroreception, they also swallow plastic items that give off no electric fields. According to other researchers, heavy objects could be ingested to solve buoyancy control problems.

As stated previously, sharks have the ability to evert their stomachs to provide a possible means to empty their stomachs when they eat indigestible objects.

On 18 April 1935, a 4.5m tiger shark was caught off Sydney, Australia. The shark was still alive, and was transported to the Coogee Aquarium for exhibition. For many days the shark fed regularly. A week after capture, the shark refused to feed and became very nervous. It began bumping into the tank walls, and after 20 minutes it regurgitated a human arm. On the forearm was a tattoo of two boxers. The shark became sick two days after regurgitating the arm, and the aquarium staff killed it. Its stomach contained part of another shark and fish

A great white shark (*Carcharodon carcharias*) approaches a shark cage. Sharks are attracted to metals as a response to the galvanic currents produced by electrochemical interactions between sea water and metals (photograph by Vittorio Gabriotti).

bones. The police asked shark specialist Victor Coppleson to examine the arm. The researcher concluded that the arm had not been bitten from the body by the tiger shark. The police issued a description of the arm, and a man identified it as that of his brother. During a series of murder trials, it was hypothesised that the victim had been dismembered, and sunk to the sea bottom in a tin trunk. The arm did not fit, so it was roped to a heavy weight and dumped in the sea, where the tiger shark ingested it.

To avoid the danger of submarines colliding with a ship when surfacing, United States Navy submarines are equipped with a listening device to detect ships. Some years ago it was found that many of the rubber coatings on these devices were damaged by strange precise slashes, which were caused by cookiecutter sharks (*Isistius brasiliensis*). Consequently, the United States Navy covered the rubber coating with a layer of fibreglass to protect it from cookiecutter shark bites.

PREY DEFENCES AGAINST PREDATORS

The reaction of marine animals to the presence of sharks varies with species and circumstance. Sharks are the sea's most feared predators, and their potential prey have developed several safety measures to defend themselves against the danger.

Natural selection has acted to reduce the probability of shark attacks. All prey have developed behaviours, and morphological and physiological characteristics, that significantly minimise the risk of shark attacks.

DEFENCE STRATEGIES

Defensive strategies of the prey impose strong deterrents to sharks and other predators, and in some cases render the prey extremely difficult to capture. Some marine animals are capable of inflicting serious injuries to predatory sharks, and many sharks bear scars and scrapes from fights with prey. For example, many animals swim in groups rather than alone. This behaviour might provide some margin of safety to an individual, by reducing the probability of capture by predators such as sharks. A group of potential prey would be more likely than a solitary animal to detect an approaching shark and to take evasive action (see 'Hunting aggregated prey', page 140). For example, great white sharks (*Carcharodon carcharias*) usually attack solitary marine mammals. Researchers also report that isolated human beings are more frequently attacked than those in groups.

Fish and elephant seals often swim along the sea bottom, very close to the substrate. The potential prey has a view of what is above and ahead, and has a higher probability of seeing the predator before it is seen and attacked. Increased visibility of predators positioned overhead constitutes a great benefit. On ascent from deep waters, marine animals scan the near surface zone for sharks and other predators. If a dangerous animal is sighted, they can change direction or return to deeper zones.

Many adult male seals and sea lions sustain wounds during aggressive competition for mates, and these animals usually avoid entering the water when freshly wounded or bleeding.

Prey can learn how to avoid shark predation. Reiji Masuda and David A Ziemann have studied post-release mortality in stock enhancement projects. They released Pacific threadfins (*Polydactylus sexfilis*) into experimental tanks with scalloped hammerheads (*Sphyrna lewini*). Heavy predation occurred only in the first hour after the release, suggesting that Pacific threadfins learned how to avoid hammerhead predation in a relatively short period.

Echinoderms have natural defences against sharks. Sea urchins are equipped with long sharp spines composed of calcium carbonate, and use this weapon to defend themselves. The sea cucumber (*Actynopiga agassizi*) is an echinoderm, is closely related to starfish and sea urchins, and is able to defend itself against shark attacks. When a shark tries to swallow a sea cucumber, this animal is quickly ejected from the mouth. The sea cucumber produces a natural shark repellant, a toxic secretion called holothurin. The shark smell and taste receptors are affected by the holothurin. This toxic secretion is sufficient to discourage sharks.

A great white shark (*Carcharodon carcharias*) approaches a school of small fish. Swimming in a group rather than alone reduces the probability of capture by sharks and other predators (photograph by Vittorio Gabriotti).

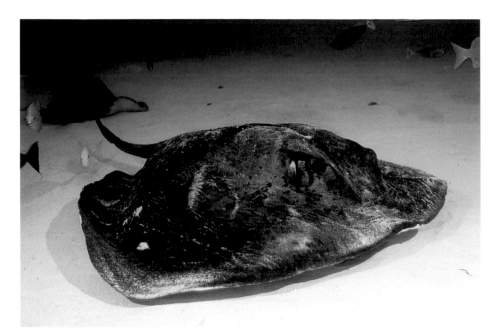

The blotched fantail ray (*Taeniura meyeni*) is a large stingray attaining lengths of up to 3.3 metres. These graceful cartilaginous fish have a venomous spine on the tail that is used for defence against predators (photograph by Vittorio Gabriotti).

Crustaceans are armoured, heavy-shelled or equipped with defensive armaments such as spines and claws. The pinch of large and powerful claws can be extremely painful.

The swordfish (*Xiphias gladius*) has a powerful weapon against sharks: a long, pointed snout shaped like a hard sword, which is roughly one-third of its body length. Large swordfish are vulnerable when asleep basking in the sun. Otherwise, they are formidable opponents, capable of inflicting severe wounds on a potential attacker. The sharks that prey on swordfish, such as the shortfin mako (*Isurus oxyrinchus*) and great white sharks, have a difficult time capturing these large predators.

The seacatfish (family Ariidae) have venomous spines on the first dorsal and pectoral fins. These spines are strongly serrated, and are covered with toxic mucous tissue. Nevertheless, seacatfish are among the most common prey items of the great hammerhead (*Sphyrna mokarran*) and bull shark (*Carcharhinus leucas*). David A Ebert examined a broadnose sevengill shark (*Notorynchus cepedianus*) containing the spine of a seacatfish that had penetrated through the roof of the shark's mouth and into the cranium.

The Moses sole (*Pardachirus marmoratus*) inhabits the Red Sea. This fish produces a natural shark repellant, a toxic secretion contained in the protective mucus covering its skin. The shark olfaction and gustatory receptors are adversely affected by this secretion. Consequently, the predator avoids biting the Moses sole, whose repellant is sufficient to protect this bony fish from being eaten by sharks.

RAYS

Stingrays (family Dasyatidae) have a long venomous spine on the tail that is used for defence against predators. These spines can cause excruciating pain and death to humans. Usually this weapon provides some margin of safety from shark predation, but it does not seem to be feared by the great hammerheads, lemon sharks (*Negaprion brevirostris*) and bull sharks (*Carcharhinus leucas*). Stingray spines are often found in the mouths of great hammerheads and lemon sharks. Shark specialist Perry Gilbert reported one large hammerhead that was found to have 96 stingray spines in its mouth and head. American researcher Guido Dingerkus observed several lemon sharks with stingray spines embedded in their jaws. Geremy Cliff and Sheldon FJ Dudley found a stingray spine loose in the coelomic cavity of a bull shark caught off KwaZulu-Natal, South Africa; the spine had presumably passed through the wall of the digestive tract.

Torpedo rays have electric organs that produce a high-voltage electric current which can be used to shock and capture their prey. These electric organs are also a weapon against predators. In South African waters, David A Ebert examined an Atlantic torpedo ray (*Torpedo nobiliana*), with a disc width of 31.5cm, that had a shark scar. A bluntnose sixgill shark (*Hexanchus griseus*) had apparently seized the torpedo ray exactly on its electric organs and had received a shock (the Atlantic torpedo ray is able to produce an electrical potential of up to 220 volts).

DOLPHINS

The reaction of dolphins to the arrival of sharks varies greatly, depending on certain factors and circumstances. In captivity, bottlenose dolphins (*Tursiops truncatus*) trained to repel sandbar sharks (*Carcharhinus plumbeus*), lemon sharks and nurse sharks (*Ginglymostoma cirratum*) became nervous in the presence of bull sharks. In rare cases sharks have been killed by dolphins when the animals were together in captivity. Bottlenose dolphins aggressively chased tiger sharks (*Galeocerdo cuvier*) in the tank, killing two young individuals.

In nature, reported reactions of dolphins to the presence of these predators vary from ignoring the shark, to herding it, to avoiding it. However, in general dolphins avoid close encounters with large sharks. Photographer Marty Snyderman observed a dozen bottlenose dolphins (*Tursiops truncatus*) and several dusky sharks (*Carcharhinus obscurus*) swimming side by side without any incident as they fed on flying fish (family Exocoetidae) during night-time at Socorro Island, Mexico. Such occasions, however, appear to be more the exception than the rule. Dolphins avoid shark predation by staying in groups. In South African waters, CK Taylor and GS Saayman observed a group of bottlenose dolphins avoiding a large hammerhead shark (*Sphyrna* sp.). The cetaceans divided into two smaller groups and increased speed, then they swam past the hammerhead shark on either side before rejoining.

A report of avoidance reaction by a group of nine bottlenose dolphins to the approach of a great white shark in Shark Bay, Western Australia, has been described by Richard C Connor and Michael R Heithaus. They observed a 2.5-3m great white shark traveling slowly with its dorsal fin protruding above the surface. The shark turned toward a group of eight to nine dolphins, all females

A group of spinner dolphins
(*Stenella longirostris*) (photograph
by Vittorio Gabriotti)

A group of Australian fur seals (*Arctocephalus pusillus doriferus*). These pinnipeds often occur in groups and, with their speed and manoeuverability, are difficult to capture (photograph by Vittorio Gabriotti).

and calves, floating at the surface about two hundred metres away. The predator did not deviate from its course until it swam into the dolphin group. Dolphins continued to float at the surface until the shark approached to within 2-3m, when the group of dolphins suddenly submerged, then emerged leaping. The leaping dolphins coalesced into two groups, one group of four individuals leaping in a north-westerly direction and one group of five individuals leaping in a north-easterly direction. The dolphins ceased leaping some minutes after they began, and the two groups rejoined 27 minutes after the shark encounter. None of the dolphins exhibited any evidence of having been bitten by the white shark.

In another case recorded in Shark Bay, two groups of bottlenose dolphins traveling slowly joined to form a group of 15 individuals including two calves. The group began swimming fast. Then a mother with calf that had been foraging 30 metres away joined the group. Just as the group began moving away fast, a shark of unknown species moved away from the direction in which the dolphins were traveling. After a few minutes the cetaceans slowed down, and the mother and calf that had been foraging left the group.

In another case, a group of bottlenose dolphins dived to more than ten metres when they encountered a great white shark.

SEALS AND SEA LIONS

Sea lions are extremely agile. Undoubtedly, these pinnipeds are able to escape shark attacks. Several times, underwater photographer Walter Heim has encountered California sea lions (*Zalophus californianus*) while chumming for shortfin mako sharks (*Isurus oxyrinchus*) off San Diego, California. The sea lions were aggressive, and chased away mako sharks of similar size. Underwater photographer Marty Snyderman has often observed California sea lions competing with blue sharks (*Prionace glauca*) and shortfin mako sharks for bait. Sea lions stole the bait from the sharks several times. Moreover, the sea lions moved behind and above the shark and bit it on the caudal peduncle. Snyderman reported that several sharks became so annoyed at being bitten by the pinnipeds that they departed. Such behaviour may seem like play, and similar movements are exhibited when sea lions meet scuba divers and perform acrobatics underwater, moving behind one of them to pick at their flippers. This behaviour serves as valuable practice for future shark encounters.

Wesley Rocky Strong, Rodney Fox, and Ron and Valerie Taylor observed great white sharks being harassed by Australian sea lions (*Neophoca cinerea*) and New Zealand fur seals (*Arctocephalus forsteri*).

Burney J le Boeuf and Daniel E Crocker studied the diving pattern of elephant seals, and concluded that the movement of these pinnipeds is, in part, a behavioural adaptation for avoiding encounters with near-surface predators. Behaviours for minimising encounters with great white sharks include long-duration horizontal deep dives for ten to 22 minutes in the depth range of 200-600m, with brief interdive surface intervals for one to three minutes. These animals simply dive as deep as possible and avoid swimming at the surface.

Northern elephant seals also reduce the probability of encountering great white sharks by swimming faster in waters over the continental shelf than off it, swimming at the sea bottom, surfacing for brief intervals, and minimising splashing. Moreover, their newly weaned pups learn to swim close to shore.

All these behaviours provide some margin of safety from shark encounters. Otariid diving patterns have been observed, and do not exhibit these particular behaviours associated with avoidance of near-surface sharks.

An Australian fur seal (*Arctocephalus pusillus doriferus*)
plays with a diver, moving behind him. A similar behaviour
is exhibited when these marine mammals meet sharks
(photograph by Vittorio Gabriotti).

SHARK PREDATORY TACTICS: INTRODUCTION

To circumvent the defences of prey, sharks rely on their own predatory tactics. They need to be able to capture prey as efficiently as possible, and in fact behavioural adaptations do indeed allow these predators to catch and feed efficiently. Many sharks are superb predators.

The sections that follow deal with different shark predatory tactics. Since evolution depends upon survival of the fittest, predatory success plays a very important role in shark survival. These formidable predators have evolved a wide variety of predatory strategies including biting, grasping, gouging, suction and filter feeding. The success of many sharks depends on speed and the element of surprise.

The strategy classifications featured in this book are basically the same as those proposed by Scott W Michael in his book *Reef sharks and rays of the world*: we shall therefore discuss pursuit hunters, stalking hunters, ambush hunters, hunters of hidden prey, predators on armoured animals, predators on aggregated prey, scavengers, hunters of prey out of the water, and cooperative hunters.

Sharks are versatile predators, and no shark relies exclusively on a single tactic to capture its prey. For example, the tiger shark (*Galeocerdo cuvier*) is a scavenger but also a predator of live animals, and is able to capture large invertebrates and fish, sea birds at the sea surface, and marine turtles in shallow water near the beach.

Sharks learn from past experience and are able to refine their predatory skills. They improve their hunting abilities and test them. Sometimes great white sharks (*Carcharodon carcharias*) show a particular behaviour described by Canadian shark expert R Aidan Martin: having caught a pinniped, the predator tossed it into the air and abandoned it. Killer whales (*Orcinus orca*) show a behaviour very similar to that of great white sharks. R Aidan Martin suggested that tossing may be a form of play or a hunting practice (see also 'Shark attacks on humans', page 82).

Great white sharks (*Carcharodon carcharias*) learn from past experience and are able to refine their predatory skills. They improve their hunting abilities and test them (photograph by Vittorio Gabriotti).

PURSUIT HUNTERS

Many active sharks pursue and catch their prey; numerous sharks regularly employ this foraging strategy. Since these predators feed on many fast-swimming creatures, they need to be able to accelerate rapidly and make fast and agile movements to catch rapid prey. Powerful strokes of the tail propel them through the water at a high speed. Some of these sharks are characterised by modifications related to muscular power and speed in swimming, like a highly streamlined body, conical snout, strong horizontally flattened caudal keels, lunate caudal fin, and the ability to maintain a body temperature that is higher than the ambient water.

The blue shark (*Prionace glauca*) is able to accelerate rapidly and make fast and agile movements in pursuit of its prey (photograph by Walter Heim).

The shortfin mako (*Isurus oxyrinchus*), great white shark (*Carcharodon carcharias*), porbeagle (*Lamna nasus*), salmon shark (*Lamna ditropis*) and thresher sharks (*Alopias* sp.) are warm-bodied, because they have heat-retaining systems. An increase in temperature makes it possible for more energy conversion to work, thus enabling these sharks to swim faster (see also 'Rate of food consumption', page 46). Consequently, the elevated body temperature enables them to pursue and capture fast prey. These fast sharks swim with the body almost rigid, with powerful strokes of the caudal fin, and they are astonishingly athletic.

Sharks usually swim at relatively slow speeds. The slow movements are deceptive, because they conserve energy and reduce the likelihood of alerting prey. Speed bursts are saved for feeding.

The shortfin mako, grey reef shark (*Carcharhinus amblyrhynchos*), blacktip reef shark (*Carcharhinus melanopterus*), broadnose sevengill shark (*Notorynchus cepedianus*), certain smooth-hounds (*Mustelus* sp.) and hammerhead sharks (*Sphyrna* sp.) are pursuit hunters. These predators concentrate on a prey and pursue it. Even the apparently slow bluntnose sixgill shark (*Hexanchus griseus*) is a pursuit predator, able to capture large fast-swimming animals such as cetaceans, swordfish (*Xiphias gladius*) and dolphinfish (*Coryphaena* sp.). The frilled shark (*Chlamydoselachus anguineus*), one of the most unusual of all sharks, may actually be included among the pursuit predators since its diet includes epipelagic squids with strong swimming capabilities, such as the boreal clubhook squid (*Onychoteuthis borealijaponica*), purpleback squid (*Sthenoteuthis oualaniensis*) and Japanese common squid (*Todarodes pacificus*). However, Tadashi Kubota and colleagues have suggested that frilled sharks may feed on injured squids and specimens that sink owing to exhaustion after spawning.

Some sharks, such as the shortfin mako, great white shark and requiem sharks (family Carcharhinidae), have been observed pursuing prey animals and incapacitating them by biting off their tails. At the Milan fish market, Italy, one of Europe's largest markets, veterinarian Luigi Piscitelli has often observed swordfish bearing healed scars inflicted by shark bites. He reported that most shark bite scars were distributed over the posterior part of the body, caudal peduncle and caudal fin. These wounds are the tangible evidence of predatory attempts on these fast-swimming fish rather than post-mortem scavenging. In aquaria, requiem sharks (family Carcharhinidae) have been observed incapacitating bony fish by biting off their tails before eating them.

A 400-million-year-old extinct shark, *Cladoselache*, belonging to a group called cladodonts, reached lengths up to 2 m, and had a terminal mouth, long jaws with multi-cusped teeth, stout dorsal fin spines, and a caudal fin similar in shape to those of fast-swimming mackerel sharks (family Lamnidae), in which the upper and lower lobes are almost equal. It was a pursuit predator, and some

The shortfin mako (*Isurus oxyrinchus*) is the fastest shark and one of the fastest creatures in the sea, with an estimated speed of 35-56 kilometres per hour (photograph by Walter Heim).

fossilised specimens contained whole fish swallowed tail first. Its jaws and teeth were adapted to seizing the prey and swallowing it whole.

The shortfin mako has a strongly conical snout, wide caudal keels, a lunate caudal fin (symmetric lobes), and long curved teeth. The mako is the fastest shark and one of the fastest fish (some think the fastest creature in the sea), with a speed of 35-56kmh. This high speed is a result of this shark's ability to maintain a body temperature that is higher than the ambient water. Shortfin makos have the speed and agility to catch mackerels, herrings and tunas. A 180kg mako shark can capture and eat 5.4-6.8kg bluefish (*Pomatomus saltatrix*) whole or in two to three pieces.

The speed of shortfin makos also enables them to pursue and attack powerful swordfish. A swordfish of approximately 1m severed in five pieces was found by Luigi Piscitelli in the stomach of a young shortfin mako that was about the same size as its prey. An almost-intact 50kg swordfish was found inside a 327kg shortfin mako caught at Bimini, Bahamas. Another shortfin mako with an almost whole swordfish in its stomach showed ten to 12 wounds in its skin that were attributed to fighting with the powerful prey. An estimated 60-80kg shortfin mako was witnessed pursuing a swordfish inside a tuna trap off Bonagìa, Sicily, Italy. The following day both mako and swordfish were found trapped in the nets; the mako was netted at a distance of one metre from its prey.

Predation by shortfin makos on that kind of dangerous fish can be a reason for serious injuries. An adult female shortfin mako caught off KwaZulu-Natal, South Africa, had an 18cm-long bill of a small Indo-Pacific sailfish (*Istiophorus platypterus*) embedded in its left orbit. The lack of vision in this eye may have impeded feeding, as the shark appeared to have lost much weight. Sometimes large mako sharks also chase dolphins. The presence of a dolphin caudal fin in the stomach of a 3.9m shortfin mako caught off Bagnara Calabra, Italy, suggests that these sharks incapacitate the small cetaceans by biting off their tails.

Great white sharks are capable of pursuing and capturing fast prey. Researchers observed great white sharks as they made high-speed chases of pinnipeds for

An Indo-Pacific sailfish (*Istiophorus platypterus*). The shortfin mako (*Isurus oxyrinchus*) often attacks billfish (photograph by Shawn Dick / Aquatic Release Conservation / NOAA Fisheries Service, Southeast Fisheries Science Center).

distances of more than one hundred metres. The association between tunas and white sharks near tuna traps has often been reported by ichthyologists, particularly from Sicilian waters and the eastern Adriatic Sea. Predation by great white sharks on tuna is the reason for numerous simultaneous captures, because these great white sharks were caught in tuna traps while they were pursuing their prey.

Examination of great white shark stomach contents revealed that dolphins are an important item in the diet of this species. Great white sharks not only scavenge on carcasses of these odontocetes, but also catch free-swimming individuals. Great white sharks probably execute attacks on live dolphins by biting the caudal part of the cetacean. A great white shark can easily sever a dolphin caudal peduncle with a single bite. This initial bite incapacitates the dolphin because the posterior part of the body is the region responsible for locomotion. In zones where pinnipeds are absent or sporadic, dolphins seem to replace pinniped importance in the diet of great white sharks. A good example is the Mediterranean Sea, as is evidenced by numerous cases collected in the Italian Great White Shark Data Bank, held by the author. Other predatory tactics, like

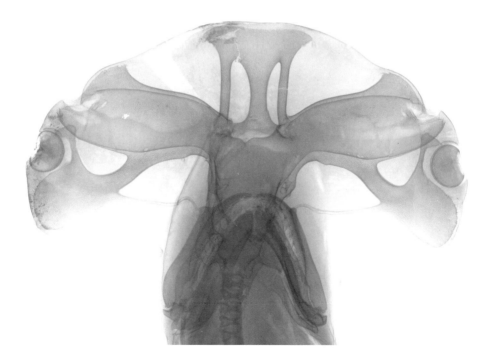

Radiography of the head of a scalloped bonnethead (*Sphyrna corona*) from Panama, Pacific Ocean (photo courtesy of the California Academy of Sciences Department of Ichthyology)

a vertical approach, may be used by great white sharks to attack these fast-swimming animals (see 'Stalking hunters', page 116).

Even thresher sharks (family Alopiidae) rely on speed and agility to catch their prey. The thresher shark caudal fin is strongly asymmetrical. Its upper lobe is almost as long as the rest of the body, and the vertebrae in the apex of the tail have expanded dorsal and ventral processes. A very interesting observation concerning thresher shark predation was made by researcher WE Allen in California waters. He observed a common thresher shark (*Alopias vulpinus*) of about 1.8m pursuing a small bony fish, possibly a longfin smelt (*Spirinchus thaleichthys*). The shark passed partly ahead of the prey, then turned quickly and gave a pair of whip strokes with the incredibly long upper lobe of its caudal fin. The small fish was injured and died shortly after being hit, but the common thresher shark fled, frightened perhaps by the presence of the observer. Sport fishermen off southern California fish the thresher shark in the spring by trolling baited lures, dead baits and live baits. About half the sharks are caught tail-hooked. Many times, a live bait is swatted by the thresher tail, which crushes the bait fish. The implication is that these incredible fish can pursue a preyfish and simultaneously use the tail as a weapon on a single targeted fish. It has been suggested that they use the leading edge of the tail rather than the side. A hooked thresher shark next to the boat is capable of inflicting damage on the fisherman.

Dusky sharks (*Carcharhinus obscurus*) have been observed catching flyingfish (family Exocoetidae) at Socorro Island, Mexico. Photographer Marty Snyderman eyewitnessed and described their predatory tactic. A dusky shark singles out a flyingfish and swims at it from behind, until the prey is only about 30cm in front of its snout. Then the flyingfish tries to move off to the side, but the shark turns its head to the same side to capture the prey. All observed predatory attempts were successful.

Hammerhead sharks have evolved particular and interesting predatory strategies. The head is greatly flattened and expanded sideways, so as to resemble a large hammer. The hammer-shaped head enables hammerheads (family Sphyrnidae) to catch fast prey, because one function of the wide head is improved hydrodynamics. Other functions of the hammer are spreading the ampullae of Lorenzini and the lateral line system over a wider area, and spacing eyes and nostrils farther apart in order to increase mechanoreception, electroreception, chemoreception and photoreception. Moreover, the head plays an important role in the handling of rays, which are a favourite food for hammerheads. These sharks use the sophisticated sensory capabilities of their wide heads to detect rays buried under the sand (see Chapter 18: 'Searching for hidden prey').

The principal prey of the great hammerhead (*Sphyrna mokarran*) is stingrays. Two detailed observations of batoid prey-catching by the great hammerhead in the waters of Bimini, Bahamas, have been reported by Wesley R Strong, Demian D Chapman and Samuel H Gruber. The attacks occurred on a southern stingray (*Dasyatis americana*) and a spotted eagle ray (*Aetobatus narinari*). In both cases, an initial hammerhead bite removed most of one of the ray's pectoral fins, which disabled its propulsive mechanism. In both cases, the hammerhead pushed the mutilated ray to the sea floor and took bites from the prey body. The shark used its head to pin the prey against the bottom, while manoeuvering its body so that it could grasp the ray in its mouth. The hammerhead swam around the ray for many minutes before it finished eating the prey.

Spotted eagle rays (*Aetobatus narinari*) are prey of the great hammerhead (*Sphyrna mokarran*) and other shark species such as the shortfin mako (*Isurus oxyrinchus*) and silvertip shark (*Carcharhinus albimarginatus*) (photograph by Vittorio Gabriotti).

The shortfin mako (*Isurus oxyrinchus*)
relies on speed and agility to catch its prey
(photograph by Walter Heim).

STALKING HUNTERS

Not all sharks are able to pursue and capture fast-swimming animals from behind. As noted previously, the success of many sharks depends on both speed and the element of surprise.

Stalking hunters attack the fast-swimming animals on which they feed suddenly and violently. The victim does not see the shark until it is too late. Most sharks mainly feed at twilight and night (see 'When sharks feed', page 52). Under the cover of complete darkness, prey have more difficulty detecting predator approach. Some sharks have a special preference for turbid waters. Under conditions of poor water visibility, it is easier to approach a prey unnoticed. In some sharks, the tips of the underside of the pectoral fins are black. Richard Ellis and John E McCosker have suggested that this colouration is a fine tuning of the camouflage pattern. According to these specialists, the black spot compensates for the flash of white that might otherwise alert a potential prey when the fins flex as the predator turns.

Species known to stalk their prey include the great white shark (*Carcharodon carcharias*), broadnose sevengill shark (*Notorynchus cepedianus*), wobbegongs (family Orectolobidae), nurse shark (*Ginglymostoma cirratum*), sandtiger shark (*Carcharias taurus*) and catsharks (family Scyliorhinidae).

The great white shark is a master at surprise attacks, and often uses this hunting technique. Most great white shark victims did not see the predator prior to the attack.

Italian photographer Vittorio Gabriotti described a particular approach on a possible prey performed by great white sharks at the Neptunes Islands, in the mouth of Spencer Gulf, South Australia. The large quantities of tuna oil, blood and macerated tuna used to attract the sharks to the vessel also attracted schools of jack mackerel (*Trachurus declivis*). Some great white sharks, of estimated total lengths of 3.5-4m, performed the following tactic when approaching the observer. On the first pass the shark made towards the cage, it approached from the side of the cage that was covered by numerous jack mackerels. Then the shark observed the divers, remaining at the visibility limit behind the school of fish. Vittorio Gabriotti has suggested that great white sharks may use this tactic to increase the element of surprise when attacking prey, and to estimate the size of the prey, and its strength, before executing an attack. The white shark moves towards a prey item using a tightly packed school of fish as camouflage to obscure its approach. The shark can continue to evaluate the parameters of the predatory situation and remain undetected at the visibility limit behind the

A great white shark (*Carcharodon carcharias*) makes its first pass towards the photographer, Vittorio Gabriotti, approaching him from behind a large school of jack mackerels (*Trachurus declivis*) (Fig. 1), then the predator observes the photographer remaining at the visibility limit behind the school of fish (Fig. 2) (photograph by Vittorio Gabriotti).

Australian fur seals (*Arctocephalus pusillus doriferus*).
Pinnipeds are a major food source for the great white shark
(*Carcharodon carcharias*) (photograph by Vittorio Gabriotti).

school of fish, swimming on an axis aligning the shark, school of fish, and prey. We have named this behaviour 'hidden approach'. The observation that larger white sharks did not show this kind of approach behaviour could be related to their greater size and strength. It could also be further indication of a lack of necessity for such a cautious behaviour beyond a certain point in the growth of the shark.

Pinnipeds are a major food source for great white sharks, so the hunting strategy of this predator is adapted to the life history of pinnipeds, which form colonies on islands and coastlines. These sharks carry out surprise attacks on sea lions and seals swimming at the sea surface.

Great white sharks approaching a prey can be oriented horizontally or vertically. The predator uses its heavy mass and speed to violently ram and stun the prey. During vertical approaches, the great white shark attacks its prey from below, by swimming from deep depths (up to 17m) and moving on a line that is 45°-90° oriented from the prey. The shark swims so fast that the prey can be flung out of the water. Often the prey is disoriented, and therefore incapable of resistance.

Wesley Rocky Strong Jr has underlined the advantages of attacking near-surface prey from below. The majority of approaches that he observed during a study of great white sharks in Spencer Gulf, Australia, were horizontal, but vertical approaches were also common. The predator is less visible when coming directly from the deep, and it has the best view of the prey silhouetted against the surface light. The prey has fewer escape paths as the prey is pinned against the surface (escape in the direction opposite to that of the predator is impossible). Great white shark attacks on humans can also be oriented horizontally or vertically. Usually the shark approaches from below, the side or behind its victim. Frontal attacks are rare. Wesley Rocky Strong Jr has also observed 2.2m young white sharks vertical swimming, indicating that this behaviour precedes variation in the diet. As stated previously, young white sharks measuring less than 3m in length feed primarily on fish, while larger individuals are important predators and scavengers of marine mammals.

Several studies at Southeast Farallon Islands off San Francisco, and Año Nuevo Island off Santa Cruz, California, USA, have been conducted by A Peter

The great white shark (*Carcharodon carcharias*). This formidable animal is a master at attacking by surprise. Most great white shark victims did not see the predator prior to the attack (photo by Vittorio Gabriotti).

Klimley, Peter Pyle, Scot Anderson, Sean van Sommeran, and Burney J le Boeuf and colleagues. These islands and the adjacent California coastline are the home to colonies of phocids and otariids, including California sea lions (*Zalophus californianus*), northern elephant seals (*Mirounga angustirostris*), Steller sea lions (*Eumetopias jubatus*) and harbour seals (*Phoca vitulina*). Researchers described the hunting strategy of the great white sharks.

The prey killed by a shark is not always identifiable owing to the fact that dead sea lions sink, while dead seals float. Nevertheless, seals are eaten more often than sea lions by great white sharks. This preference may reflect the fact that sea lions, with their speed and manoeuverability, are more difficult to catch, but there are other differences that may influence differential pinniped predation. Sea lions often occur in groups, while elephant seals are usually solitary when away from shore. This solitary behaviour of elephant seals makes them more vulnerable to predation.

When sea lions swim, they use their front flippers to push themselves through the water, while their rear flippers are used to help steer. Seals swim by flapping their rear flippers, while their front flippers help them steer. Pinnipeds are easily identified in that sea lions have ear flaps, which seals do not.

Great white shark attacks on elephant seals have been repeatedly observed, and the attacks follow a general pattern. The great white shark executes an initial attack on the near-surface pinniped from below or behind, usually inflicting a deep bite on the rear part of the body. Biting the rear part of the seal disables the propulsive mechanism of the pinniped. The initial attack is often followed by a waiting period. The elephant seal usually does not flee, owing to its wound or to shock, and dies from blood loss. The great white shark returns within one to five minutes to eat the dead animal. This predatory tactic enables the shark to obtain its meal with minimal risk of injury as well as minimal energy expenditure. Timothy C Tricas and John E McCosker termed this behaviour 'bite-and-spit'.

In some cases, however, the prey may escape, and survive or die far from the attack location if it is strong enough to swim. Sometimes pinnipeds drag themselves out of the water and die on the beach as a result of shark bites. This behaviour may allow many human victims the possibility to escape or be rescued. However, Australian researcher John West found that a high percentage of great white shark attacks on humans do not conform to this behaviour; consequently 'bite-and-spit' is not the rule. Even biting the hindquarters is not a rule since Douglas J Long and colleagues observed wounded and dead elephant seals with shark bite scars distributed over the entire body, including the head and neck.

In fact, sometimes great white sharks attempt to decapitate their prey. Sea lions that survive a great white shark attack have bite scars distributed predominantly on the posterior part of the body. Sea lions are often able to escape from a great

An Australian fur seal pup (*Arctocephalus pusillus doriferus*). The great white shark (*Carcharodon carcharias*) attacks young pinnipeds more frequently because they are more vulnerable than adults and have a higher fat content (photograph by Vittorio Gabriotti).

white shark attack because a bite to their rear flippers does not incapacitate them. The great white shark executes a successful attack on a sea lion usually by inflicting a bite on the mid part of the body.

Young pinnipeds are attacked more frequently than adults. Young pinnipeds are the target of attacks more frequently because they have a higher fat content, are less vigilant, have less experience with sharks than do adults, and may behave in a manner that makes them more vulnerable to predation. In fact, immature pinnipeds are observed with great white shark bite scars more often than adults. Great white sharks have a special preference for young elephant seals one to two years of age (from 1.4 to 1.7m in length). At the Farallon Islands, great white sharks kill more seals in the fall. Similarly, at Año Nuevo Island, shark bitten seals are more common during the fall and winter. At this time, young elephant seals are more abundant at these sites.

The great white shark prefers coastal waters with a median depth of 20 metres because the density of juvenile seals is highest close to shore. The area patrolled by the great white sharks at Southeast Farallon Islands and Año Nuevo Island extends from a few metres to 1.3km offshore and surrounds the pinniped colonies, where they would have the greatest chance of capturing prey. Within this area, predation is most frequent adjacent to rookeries and beaches where pinniped colonies concentrate, and particularly near entry and departure points of pinnipeds. Great white sharks tagged and monitored at Año Nuevo Island rarely ventured far from shore, but stayed very close and at times approached to within two metres of the shore. At Año Nuevo Island, the sharks visited the seal colony area every day and patrolled this zone both during the daytime and at night. Some of the sharks moved back and forth parallel to the shoreline, 200-300 metres from shore, where they were ideally positioned to intercept and stalk seals and sea lions departing from and returning to the rookeries. Tidal height and other factors also influence shark predation on pinnipeds (see 'When sharks feed', page 52).

As stated previously (see 'Pursuit hunters', page 108), some sharks have been reported to incapacitate prey by biting off their tails. A vertical approach may be commonly used by these predators to obtain this result when pursuing fast-swimming animals, such as tuna and dolphin. A single bite on the caudal peduncle of these large prey can sever swimming muscles, the spinal column and blood vessels, thereby immobilising the prey. In these cases, the shortfin mako (*Isurus oxyrinchus*) and great white shark swim at greater depth than their prey, giving them a view of what is above and a higher probability of seeing and attacking the prey, via a rapid vertical approach, before being seen themselves. This predatory tactic explains the presence of wounds over the posterior part of the body and caudal peduncle inflicted on some bluefin tuna by both great white sharks and makos.

In order to avoid detection by dolphins and porpoises, great white sharks approach these mammals from below, above or behind, because odontocetes have an anteriorly directed sonar and a lateral visual field. Douglas J Long and Robert E Jones examined numerous small cetaceans with wounds on the caudal peduncle, urogenital region, abdominal area, and dorsum, while wounds to the head and flanks were less common. PW Arnold reported the case of an estimated 5.2m great white shark that contained three harbour porpoises (*Phocoena phocoena*) in which all three specimens had the tail stock severed. Urogenital and abdominal regions are vulnerable areas. Shark bite scars are more often observed on the dorsum of live cetaceans since this body region is less vulnerable, providing higher probabilities of surviving the attack.

Common dolphins (*Delphinus delphis*). In order to avoid detection by dolphins, great white sharks (*Carcharodon carcharias*) and shortfin makos (*Isurus oxyrinchus*) approach these mammals from below, above or behind, because odontocetes have an anteriorly directed sonar and a lateral visual field (photograph by Vittorio Gabriotti).

AMBUSH HUNTERS

Like the stalking hunters, ambush hunters attack by surprise. While in the case of stalking hunters the victim does not see the shark until it is too late, in the case of ambush hunters the victim sees the shark but does not understand that it is a shark until it is too late.

All ambush hunters need to attract prey into their strike zone. Depending on species, ambush hunters can rely on camouflages or lures to catch their prey.

Some ambush hunters are bottom-dwelling species. They remain motionless and lie in wait for passing prey, spending most of their time on the sea floor. These

The Pacific angelshark (*Squatina californica*). The flattened body of the angelshark is an adaptation to life on the sea bottom (photograph by Tony Chess).

animals rely on camouflage to catch their prey. Many have cryptic colouration, showing complex pigmentation; a typical example is the markings possessed by the angelsharks (family Squatinidae). A few species have skin appendages that serve to break up their outline against the sea bottom. Many of these sharks are quite colourful, in contrast to most of the other species, which tend to range in colour from grey to brown. Many ambush predators have a dorso-ventrally flattened shape. Their particular shape, skin appendages and bright colour make them resemble the weed-covered rocks and coral where they live. As a consequence, these predators can be difficult to see against the sea bottom.

Most of these species do not actively hunt, but simply wait for a prey to get close to them. Sometimes it is very difficult to see these predators as they lie motionless on the sea bottom. In order to capture prey they are able to strike very rapidly. When its prey comes close enough, the ambush predator darts out and seizes it. Then the shark settles back to wait for more prey.

Well-known ambush predators are the wobbegongs (family Orectolobidae). Wobbegongs are provided with cryptic colour, including spots, blotches, stripes and irregular reticulations, that camouflage them among the rocks, algae and corals. The lips and sides of the head bear weed-like flaps of skin.

There is little possibility of confusing angelsharks with another shark, as the greatly flattened body is very distinctive. They resemble the rays in their body shape and in having very wide pectoral fins. The flattened body of the angelshark is an adaptation to life on the sea bottom. These sharks bury themselves in the sand of the sea floor, from which they strike at high speed to capture fish, squid, crustaceans and other animals, with their wide and highly protrusible jaws. The strike zone of the Pacific angelshark (*Squatina californica*) is 4-15cm from the snout. When a prey comes within the strike zone, the Pacific angelshark rarely misses.

The swellshark (*Cephaloscyllium ventriosum*) is an ambush hunter that favours blacksmith (*Chromis punctipinnis*). There are two kinds of swellshark predatory tactics used to catch blacksmiths, called 'gulp' and 'yawn'. Both techniques are characterised by the shark initially waiting for prey on the sandy sea floor near a reef. The predator remains motionless until a blacksmith comes close enough. When the small fish approaches the swellshark, the predator rapidly opens its mouth and expands its buccal cavity and pharynx to suck in the prey ('gulp'), or slowly opens its mouth and simply waits until the blacksmith swims into its open mouth ('yawn').

Some ambush hunters rely on a particular kind of camouflage to catch their prey. These sharks attract prey by mimicking their natural hiding place, such as a rock. The nurse shark (*Ginglymostoma cirratum*) uses this interesting technique to capture benthic fish, molluscs and crustaceans. Juvenile nurse sharks have been observed to remain motionless with their snout pointed upward and their body supported off the sea bottom on their pectoral fins. A potential prey may mistake the space under the body of the shark for a place to hide, and move straight toward its death.

David A Ebert has observed leopard catsharks (*Poroderma pantherinum*) and broadnose sevengill sharks (*Notorynchus cepedianus*) alter their ground colour over short time periods. The author has hypothesised that this physiological capacity may also exist in the bluntnose sixgill shark (*Hexanchus griseus*). This characteristic is advantageous for these animals to camouflage themselves as they move from one habitat to another.

Some ambush predators have lures for attracting prey. Anatomically, some species are admirably adapted to capture prey using parts of their body as bait. Others use particular behaviours to provide deceptive information to prey, thereby inducing them to approach close enough for the predator to capture them. The wobbegongs have dermal flaps and skin that resemble algae and encrusting invertebrates. Fish and crustaceans that feed on this kind of small benthic creature mistake the wobbegong body for food, and come within the strike zone.

One of best examples of an ambush hunter that uses lures for attracting prey is the oceanic whitetip shark (*Carcharhinus longimanus*). This shark preys on many fast-swimming animals, such as tuna (*Thunnus* sp.), skipjack tuna (*Katsuwonus pelamis*), mackerel (*Scomber* sp.), common dolphinfish (*Coryphaena hippurus*) and Atlantic white marlin (*Tetrapturus albidus*). It is unlikely that this predator would be able to catch such fast-swimming animals from behind.

Arthur A Myrberg Jr has formulated an interesting theory on the predatory tactic of the oceanic whitetip shark. According to this theory, these sharks provide deceptive information to schooling prey in order to attract them into the strike zone. This shark has conspicuous white spots at the apex of the fin. These white spots might be lures for attracting fast-swimming prey. Oceanic whitetip sharks usually swim very slowly. When one of these predators swims at the limit of visibility, the eyes of its prey are constantly drawn to their white-tipped fins, with the result that the greyish, countershaded body of the shark becomes indistinct, making the shark seem to disappear. So, the attention of the prey focuses on the white fin spots moving through the blue water. When two oceanic whitetip sharks move close together at such distances, the group of white spots can be seen moving in close formation. The visual effect is particularly striking during periods of low light. The white spots may appear like a school of small fish to the prey. When the prey approaches the shark, the large predator quickly accelerates and captures it. This tactic may explain how this shark is able to catch many fast-swimming creatures. The oceanic whitetip sharks are often

The oceanic whitetip shark (*Carcharhinus longimanus*). According to the theory of Arthur A Myrberg Jr, the wide white spots at the apex of the fin serve to provide deceptive information to schooling fish and thereby attract them into the strike zone (photograph by Joost Wenderich).

accompanied by many pilot fish (*Naucrates ductor*), and their presence could help to increase the particular visual effect and attract the prey into the strike zone.

Some deep-water sharks are equipped with bioluminescent organs for attracting prey. Bioluminescence is the capacity of some organisms to emit light. This phenomenon is produced by a chemical reaction between the protein, luciferin and oxygen in the presence of the enzyme luciferase. The photophores are bioluminescent organs located in the skin, usually in longitudinal rows, where light is produced by specialised cells and reflected through a lens. The degree of bioluminescence seems to vary from individual to individual.

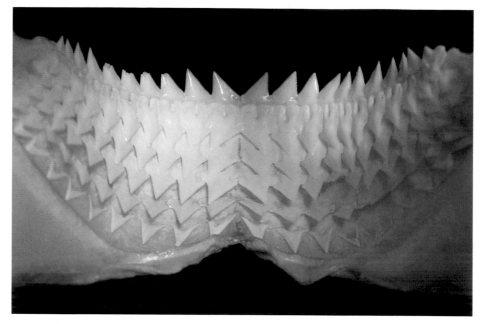

Teeth of the kitefin shark (*Dalatias licha*). Kitefin sharks may be classified as facultative ectoparasites (photograph by Alessandro De Maddalena).

The cookiecutter shark (*Isistius brasiliensis*) uses its glowing caudal fin to attract cetaceans, elephant seals (*Mirounga* sp.), tuna (*Tuna* sp.), marlins (*Makaira* sp. and *Tetrapturus* sp.), basking sharks (*Cetorhinus maximus*), whale sharks (*Rhincodon typus*) and megamouth sharks (*Megachasma pelagios*). The caudal fin has photophores that give off a vivid greenish light and attract the large animals by mimicking similar photophore organs present in their natural food sources, such as deep-water fish and planktonic organisms. When a large animal approaches a cookiecutter shark, the small predator accelerates rapidly and strikes with its powerful suctorial lips and muscular pharynx, which it uses to attach itself to the prey. The shark sinks its teeth into the flesh using its narrow, small upper teeth adapted for the purpose. The lower teeth are broad, long and fused together, forming a continuous high band that serves to excise a conical plug of flesh, leaving a rounded wound on the victim. The largetooth cookiecutter shark (*Isistius plutodus*) uses the same technique and is capable of excising more elongated plugs of flesh from the victim. Since cookiecutter sharks obtain sustenance both from external parts of larger organisms and from whole smaller prey, they have been classified as facultative ectoparasites.

Chunks of large fish are often found in the stomachs of kitefin sharks (*Dalatias licha*). These sharks may use a feeding technique similar to that of cookiecutter sharks.

128

Other species from great depths are equipped with bioluminescent organs. Sharks of the genus *Etmopterus* are called 'lanternsharks' because of the minute photophores that are distributed in specific patterns on their ventral surface. The pattern of their photophores differs between sex and among species, enabling these fish to recognise others of their species and to coordinate schooling and mating behaviours in the darkness of the deep sea. It is also possible that lanternsharks use their bioluminescence to attract their prey.

The function of the bioluminescent lining of the mouth of the megamouth shark is not known, but probably in dark waters this band may be more visible and may enable this giant to attract minute zooplankton straight into its open mouth. The beautiful colour pattern of light spots and stripes characteristic of the whale shark may possibly serve to lure zooplankton near.

Greenland sharks (*Somniosus microcephalus*) and Pacific sleeper sharks (*Somniosus pacificus*) are hosts to a copepod parasite (*Ommatokoita elongata*) that ranges in length from a few millimetres to about 8cm. The parasite firmly attaches itself to the cornea of the shark eye by an anchoring structure. Researchers have hypothesised that the relationship between these copepods and sharks may be a case of mutualism. According to this theory, these whitish-yellow copepods may induce prey to approach close enough to enable the Greenland shark and the Pacific sleeper shark to capture them. Researcher JD Borucinska and colleagues reported that these infections generally do not significantly debilitate the shark, but they also hypothesised that in some instances could lead to severe vision impairment, and possibly blindness.

SEARCHING FOR HIDDEN PREY

Numerous animals, including fish, crustaceans and molluscs, try to escape detection by hiding in sand, coral, rocky reef crevices and underwater caves. Many sharks move along the sea bottom exploring sand, crevices and holes in reefs in search of prey. Some sharks are efficient specialists in capturing hidden prey by using particular techniques or anatomical structures to excavate animals buried under sand, or to extract prey from cracks and crevices.

Researcher Adrianus Kalmijn demonstrated that sharks can find prey in the dark, even when the prey is buried in the sand, via their ampullae of Lorenzini (see 'The role of the senses in feeding', page 24). Although olfaction and electroreception are used by all sharks during hunting, these senses may be of special importance when sharks are looking for hidden prey. In fact, some of these sharks have enhanced sensory capacities in certain parts of the body. Depending on species, these sharks rely on agility, flexibility, intuition, force, olfaction, electroreception or special anatomical structures to catch their prey.

Suction evolved for the intake of water for respiration, but certain species employ their suctorial mouth for engulfing prey. This suction movement is much stronger than during respiration. Wobbegongs (family Orectolobidae), nurse sharks (family Ginglymostomatidae) and epaulette sharks (*Hemiscyllium ocellatum*) have been reported to use suction to capture their prey.

Suction in the nurse shark (*Ginglymostoma cirratum*) has been described by American researcher Philip J Motta and colleagues. This docile shark is usually found resting on the bottom, in reefs, or in underwater caves where it also searches for its prey. When the nurse shark detects prey, it places its mouth close to the animal. The nurse shark has a small mouth, almost terminal and laterally occluded, with modified labial cartilages and small teeth. The powerful stereotyped suction action is generated by rapidly and vigorously expanding the buccopharyngeal cavity. The prey is extracted from its crevice and entrained in a mass of water that is transported toward the motionless shark, into the mouth and out of the gill slits. The prey is often swallowed without being grasped by the teeth. Sometimes the shark raises itself on its pectoral fins while sucking the prey into its mouth. The suction may be accompanied by a loud popping sound. According to Australian shark expert John D Stevens, the nurse shark is able to produce a suction pressure of up to one kilogram per square centimetre, the equivalent of one atmosphere of pressure at sea level.

Some sharks have nasal flaps expanded as long nasal barbels, and use these anatomical structures to locate prey buried in sand. These species cruise in circles

A nurse shark (*Ginglymostoma cirratum* rests in an underwater cave off Cuba. This animal uses suction to extract prey from crevices and cracks, by rapidly and vigorously expanding the buccopharyngeal cavity (photograph by Claudio Perotti).

close to the sea bottom searching for prey, with their nasal barbels touching the sand. The mandarin dogfish (*Cirrhigaleus barbifer*) and the barbeled catshark (*Poroderma marleyi*) are so named for having greatly elongated nasal barbels, suggesting that they may trail them over the sand of the sea floor to find hidden invertebrates and fish.

The epaulette shark (*Hemiscyllium ocellatum*) has a long, slender and sinuous body, and muscular leg-like pectoral and pelvic fins. This pretty predator uses its fins to 'walk' on the sea bottom, clamber on the reef, and enter crevices or branching corals to capture hidden benthic animals. The epaulette shark also forces its head into the sand of the sea floor to find buried animals detected by its nasal barbels. It flips coral with its snout to expose hidden worms and crustaceans. Researcher Scott W Michael observed this shark plunge the anterior portion of its body into a crevice, turn itself over and use suction to capture a shrimp located on the roof of the crevice.

A whitetip reef shark (*Triaenodon obesus*) patrols its territory around the edges of a reef. This predator often catches animals hidden in crevices (photograph by Vittorio Gabriotti).

The zebra shark (*Stegostoma fasciatum*) cruises and clambers on the sea bottom, and can enter crevices and narrow channels to search for hidden animals. Predatory tactics of the frilled shark (*Chlamydoselachus anguineus*) are poorly known. This curious fish has an elongated body to suit its life on the rocky sea bottom, where it may catch animals hidden in crevices. Frilled sharks are able to protrude their upper jaws well in front of the short snout, thus enabling them to swallow large animals. Among sharks that occasionally try to capture animals hidden in crevices is the great white shark (*Carcharodon carcharias*). On one occasion, a large white shark attempted for more than a hour to capture a Tasmanian abalone diver hidden in a crevice. Luckily for the diver, its efforts were unsuccessful.

Other sharks use more violent techniques to capture their prey. A tawny nurse shark (*Nebrius ferrugineus*) of about 3m was observed catching a speared fish hidden under a massive coral head that was estimated to weigh almost half a ton. By violently forcing its body into the prey's hiding place and lifting the coral head, the predator was able to reach the fish and use suction to swallow its prey. Scott W Michael observed the nocturnal feeding behaviour of the whitetip reef shark (*Triaenodon obesus*) at Cocos Island, Costa Rica: the whitetip reef shark wedges its head into crevices searching for hidden animals. The predator twists and turns with its quite slender body to penetrate deeper and deeper into the prey's den. It can also break off pieces of coral during these predatory attempts.

Houndsharks (family Triakidae) have wide nostrils, and the undersurface of the head is flattened. These anatomical structures are especially important for these predators when they are looking for animals buried under the sand. Leopard sharks (*Triakis semifasciata*) feed on innkeeper worms (*Urechis caupo*). These Echiurans live in the sandy mud of bays and estuaries. Innkeeper worms burrow in the mud, excavating u-shaped tunnels with each end of this lair reaching the water above the sea bottom. Leopard sharks use suction to extract the innkeeper worm from its tunnel, and often swallow the prey without grasping it with their teeth.

The Port Jackson shark (*Heterodontus portusjacksoni*) often hunts animals buried in sand. For this reason, the Port Jackson shark has evolved complex nostrils with deep nasoral grooves that expose the maximum surface to sea water for detection of odours. This shark pumps water and sea floor sediment through its mouth and out of its gill slits to dig molluscs and crustaceans out of the sand. The predator seizes the prey with its conical and pointed teeth, and passes it back to the rear pavement-like crushing teeth.

Certain sharks have more extraordinary anatomical characteristics that allow them to find hidden animals. The sawsharks (family Pristiophoridae) are readily recognised by the greatly expanded, elongated and flattened snout. The rostrum is armed with a row of teeth on each side, resembling that of the sawfish, and is also equipped with very long rostral barbels. These curious fish use their weapon to dig for small bottom organisms by lashing the rostrum from side to side.

The hammerhead shark (family Sphyrnidae) has a greater number of ampullae of Lorenzini and a higher density of pores on the ventral surface of its wide head. This characteristic results in an increase in electroreception. In order to find concealed animals, hammerheads use the bizarre head much in the manner of a mine-sweeper. Hammerhead sharks seem to have a particular affinity for the flesh of rays (see 'Pursuit hunters', page 108). These highly electrosensitive sharks search for rays and wrasses that are buried below the surface sand of the sea floor. They swim close to the sea bottom, swinging the head from side to side, moving in a series of tight circles or turning in figure-eights to locate animals buried in the sand. As the hammerhead reaches the source of the electrical current it drops its lower jaw and scoops the hidden animal from the sand.

The goblin shark (*Mitsukurina owstoni*) may be included among the predators provided with special anatomical characteristics for finding prey buried in the sand. This strange fish has a greatly flattened, elongated snout, and highly protrusible jaws with narrow, pointed teeth. It would be interesting to know why this fish developed such an atypical head. We can hypothesise that the goblin shark may use its bladelike snout to excavate animals buried in the sand, and its protrusible jaws to strike at high speed.

This photo clearly shows the head of a scalloped hammerhead (*Sphyrna lewini*) (photograph by Alessandro De Maddalena).

A giant moray (*Gymnothorax javanicus*). Moray eels, like many other animals, try to escape detection by hiding in rocky reef crevices (photograph by Vittorio Gabriotti).

CATCHING ARMOURED ANIMALS

Many sharks eat a wide variety of prey that are armoured, heavy-shelled, or equipped with defensive armaments such as spines and claws. These animals include invertebrates like crabs, bivalves, snails and sea urchins, as well as vertebrates like sea turtles.

Some of the larger crustaceans equipped with powerful claws may be capable of inflicting serious injury on their aggressors. Consequently, sharks that prey on armoured animals have developed behaviours that significantly minimise the risk of injury. Moreover, numerous sharks that feed on these heavily armoured animals have specialised dentition, such as multi-cusped, pavement-like and highly serrated teeth to crush the exoskeletons of the hard-shelled invertebrates, including the hardest part of the vertebrate skeleton. Some of these sharks grind their food and break it into small pieces.

Examples of species with specialised dentition that is ideal for cutting up armoured prey include the tiger shark (*Galeocerdo cuvier*), great white shark (*Carcharodon carcharias*), smooth-hound (*Mustelus* sp.), bullhead shark (family Heterodontidae) and nurse shark (family Ginglymostomatidae). Some sharks do not swallow hard parts of prey, while others ingest their prey whole and regurgitate the harder parts later. Certain sharks mutilate armoured animals, ingesting only a part of the body.

Some sharks attack crustaceans equipped with powerful claws by grasping their claws and shaking them free from their body, such as in the case of the brown smooth-hound (*Mustelus henlei*) feeding on cancrid crabs.

The tiger shark (*Galeocerdo cuvier*) is equipped with large, flat, highly serrated, cockscomb-shaped teeth, with a large notch on the lateral margin, ideal for cutting up marine turtles. The large serrae are secondarily serrated (photograph by Alessandro De Maddalena).

A hawksbill turtle (*Eretmochelys imbricata*). Morphological, anatomical and behavioural adaptations allow some sharks to cut up armoured prey, such as sea turtles, crustaceans, bivalves, snails and sea urchins (photograph by Vittorio Gabriotti).

However, not all predatory attempts are successful. The author and Luigi Piscitelli had the opportunity to observe, to the best of our knowledge, the first evidence of interspecific interaction between the kitefin shark (*Dalatias licha*) and the paromola (*Paromola cuvieri*), a huge crab of the family Homolidae. Luigi Piscitelli collected a huge paromola caught in the waters of the Ligurian Sea off San Remo, Italy. This paromola was one of the largest of its species ever recorded, with a carapace length of 20cm. During the taxidermic preparation of the crustacean, a small scar was noticed on its long third left walking leg (pereiopod), located on the merus, very close to the articulation with the carpus. Following close examination of the scar, we identified the species responsible for the attack as a kitefin shark, which had bitten the paromola during a frontal attack. It appears that the hardness of the carapace and the powerful large chelipeds of the paromola were effective in dissuading the kitefin shark.

Bullhead sharks (family Heterodontidae) feed on sea urchins. These sharks have hard skin, powerful jaws, teeth arranged in a pavement formation, and eyes

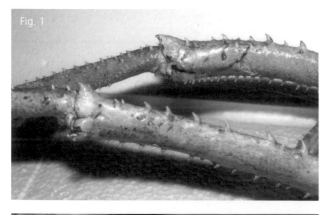

Fig. 1

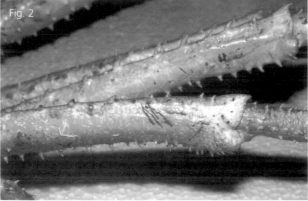

Fig. 2

Walking leg of one of the largest paromolas (*Paromola cuvieri*) ever recorded, showing a bite scar inflicted by a kitefin shark (*Dalatias licha*). Fig. 1: Anterior part of the wounded walking leg with the same scar produced by a kitefin shark's large lower teeth. Fig. 2: Posterior part of the wounded walking leg with the same scar produced by a kitefin shark's small narrow upper teeth (photographs by Luigi Piscitelli).

Set of jaws of the kitefin shark (*Dalatias licha*) (photograph by Alessandro De Maddalena).

located at the top of the head. These particular anatomical characteristics enable them to avoid possible damage caused by sea urchin spines.

In the stomachs of draughtsboard sharks (*Cephaloscyllium isabellum*), Scott W Michael discovered hermit crab remains without the shells in which these crustaceans live. Hermit crabs have a soft abdomen, and use old snail shells for protection. As the hermit crab grows in size, it must find a larger shell. The draughtsboard shark may eat the hermit crab when it is in the process of changing shells. Scott W Michael also hypothesised that the hermit crab may leave its shell when it enters the acidic environment of the shark stomach and is attacked by enzymes. The predator may regurgitate the hermit crabs and their shells to re-ingest the unprotected crustaceans.

The small-spotted catshark (*Scyliorhinus canicula*) knocks hermit crabs and snails over with its snout. The crustaceans and gastropods try to hide in their shells, but the shark seizes the prey and shakes its head to extract the animal from its shell.

Leopard sharks use an equally violent technique to eat clams. This shark bites the extended siphons of buried clams, and mutilates the molluscs by shaking its head vigorously from side to side.

The nurse shark (*Ginglymostoma cirratum*) accomplishes snail capture through suction. The snail is extracted from its shell and entrained in a mass of water that is transported rapidly into the nurse shark's mouth (see also 'Searching for hidden prey', page 130). Suction feeding is the mode of capture of the young giant clam (*Tridacna gigas*) by tawny nurse sharks (*Nebrius ferrugineus*). The spotted wobbegong (*Orectolobus maculatus*) also uses suction as its feeding method to prey on bottom invertebrates.

Teeth of the blackspotted smooth-hound (*Mustelus mediterraneus*). Smooth-hounds have specialised dentition for cutting up armoured prey (photograph by Alessandro De Maddalena).

HUNTING AGGREGATED PREY

Numerous marine animals swim in aggregations for protection or reproduction. Many bony fish, such as species of the families Clupeidae and Carangidae, swim in schools of thousands, in a tight formation. The term 'school' describes a group of animals of the same species swimming in a synchronised manner. A group of animals is more likely than a solitary individual to detect an approaching predator and take evasive action. In the face of a predator, these fish close their ranks by crowding more closely together.

Sharks often congregate near concentrated food sources, such as schools of fish and aggregations of squid. Some sharks regularly feed on schooling prey, and certain species take advantage of the migrations of the seasonal schooling fish.

A school of bigeye trevallies (*Caranx sexfasciatus*). Numerous marine animals, such as bony fish of the family Carangidae, swim in schools, in a tight formation for protection or reproduction (photograph by Vittorio Gabriotti).

The common thresher shark (*Alopias vulpinus*) (photograph by NMFS La Jolla California Shark Group)

How are small, fast-swimming fish caught by sharks? Some species employ tactics to concentrate their prey for easier feeding. Other species, such as the curious thresher sharks (family Alopiidae) and sawsharks (family Pristiophoridae), possess morphological adaptations to catch and kill small schooling prey efficiently. Yet other species have developed behavioural techniques to prey on large numbers of small animals. Blue sharks (*Prionace glauca*) and bronze whaler sharks (*Carcharhinus brachyurus*) are examples of such predators. A few sharks, such as the blue shark, whale shark (*Rhincodon typus*), basking shark (*Cetorhinus maximus*) and megamouth shark (*Megachasma pelagios*) are efficient specialists in capturing numerous small prey because they have evolved an unusual feeding method which uses the gill system for feeding as well as for respiration.

As stated previously, a few sharks are equipped with extraordinary anatomical characteristics that allow them to catch small schooling animals. The thresher sharks (family Alopiidae) are masters at feeding on schooling prey. The common thresher shark (*Alopias vulpinus*) is a strong-swimming species that frequently occurs in association with large schools of small fish. Italian researcher, Antonella Preti, and her colleagues investigated the feeding habits of the common thresher shark sampled from the California-based drift gill net fishery. They found that the northern anchovy (*Engraulis mordax*) and Pacific hake (*Merluccius productus*) are the most important species in the diet of the common thresher shark from the USA Pacific coast, followed by Pacific mackerel (*Scomber japonicus*) and Pacific sardine (*Sardinops sagax*).

As discussed previously (see 'Pursuit hunters', page 108), the thresher shark caudal fin is strongly asymmetrical because its upper lobe is almost as long as the rest of the body, and the vertebrae in its apex have expanded dorsal and ventral processes. The common name of the thresher shark is derived from the resemblance to one who thrashes, as these sharks slash the water with the incredibly long upper lobe of their caudal fin in order to herd and disorient schooling fish. Often thresher sharks circle the school and narrow the radius before attacking it. These sharks also have large eyes, enormous in the bigeye thresher (*Alopias superciliosus*), which extend onto the dorsal surface of the head. These large eyes may provide thresher sharks with a better view of the small schooling prey they are slashing.

Sawsharks (family Pristiophoridae) are another group of predators with an amazing anatomical characteristic used for capturing schooling animals (see also 'Searching for hidden prey', page 130). These sharks use their powerful rostrum, armed with a row of teeth on each side, to injure, disable or kill prey when charging into a school of fish.

A few species, including the sandtiger shark (*Carcharias taurus*), grey reef shark (*Carcharhinus amblyrhynchos*) and blacktip reef shark (*Carcharhinus melanopterus*), catch schooling fish in groups. By driving and herding their prey into shallow waters, these predators execute a simultaneous attack on the trapped fish.

Blue sharks show particular foraging behaviours when they find aggregated prey. Cephalopods form the main item in their diet in many areas. Night-time blue shark predation on a large aggregation of squid has been described by underwater explorer Philippe Cousteau, the son of Jacques-Yves Cousteau. In California waters, market squid (*Loligo opalescens*) form reproductive aggregations that soon attract blue sharks. Blue sharks do not need to concentrate on a single individual in the group; they simply swim through the dense aggregation with their jaws open, and literally fill their mouths with numerous squid. The blue shark is equipped with long gill rakers that prevent the prey from escaping through its gill slits. It also moves its head from side to side when swimming through the dense aggregation, to capture more cephalopods. An additional predatory tactic of the blue shark is a vertical attack. The predator assumes a head-up posture below the squid (an uncharacteristic feeding posture for this species) and then swims upward through the school, swallowing the squid as it

The blue shark (*Prionace glauca*) feeds heavily on cephalopods (photograph by Walter Heim)

goes. When its mouth is filled with squid, the shark sinks back down below the aggregation of cephalopods.

Other shark species, such as the leopard shark (*Triakis semifasciata*), blacktip shark (*Carcharhinus limbatus*), spinner shark (*Carcharhinus brevipinna*) and piked dogfish (*Squalus acanthias*), use similar hunting techniques characterised by swimming with mouth wide open into aggregations of prey, ingesting individuals that inadvertently are caught in their mouths. When spinner sharks and blacktip sharks catch near-surface schooling fish, they can leap spectacularly from the sea surface and fall back into the water, spinning around their body axis.

143

The whale shark (*Rhincodon typus*) may
grow to 20m, and is the world's largest
fish (photograph by Mick Jansen).

In some cases, blue sharks simply take bites from tightly massed prey. They
have been reported to use this technique to feed on tightly massed krill. Other
sharks, such as the bronze whaler shark, oceanic whitetip shark (*Carcharhinus
longimanus*) and smooth-hound (*Mustelus* sp.), use this technique to feed on
small schooling fish or krill. Bronze whaler sharks feed heavily on pilchards.
Off KwaZulu-Natal, South Africa, the South African pilchards (*Sardinops
ocellatus*) are the most common prey of the bronze whaler shark, where Geremy
Cliff and Sheldon FJ Dudley found pilchards in 84% of the stomachs of these
sharks. These selachians attack schools of pilchards, taking bites from the tightly
massed prey.

Malcolm J Smale has observed bronze whaler sharks, cape gannets (*Morus capensis*), African penguins (*Spheniscus demersus*) and skipjack tuna (*Katsuwonus pelamis*) simultaneously attacking a school of pilchards in Algoa Bay, South Africa. The African penguins swam around the school keeping the prey concentrated, the skipjack tuna propelled itself through the pilchards, the Cape gannets attacked from above, while bronze whaler sharks attacked from below, taking bites from the central part of the school. This simultaneous attack by sharks, gannets, penguins and tuna on the tightly packed school probably facilitated the activities of all the attackers. This feeding method is highly unusual for a shark, since most sharks hunt alone, are not highly social animals, and do not depend on other creatures for catching and killing prey (see also 'Cooperative hunters', page 162).

As explained previously (see 'Ambush hunters, page 124), the oceanic whitetip shark preys on many fast-swimming species, including schooling fish, by attracting these animals via the large white spots at the apex of the fin, which may appear to the prey as a school of small fish. The predator swims very slowly, but when the prey approaches the shark quickly accelerates and captures it. It has also been suggested that oceanic whitetip sharks may swim with their mouths opened wide through near-surface schooling fish when these animals are being exploited by other predators, such as tuna and mackerel. Since these other predators leap from the sea surface in pursuit of small prey, they may jump straight into the wide-open mouth of the feeding oceanic whitetip shark.

The bull shark (*Carcharhinus leucas*) has been observed to catch schools of fish in estuaries and rivers. Geremy Cliff and Sheldon FJ Dudley examined an individual that contained 186 South African pilchards (*Sardinops ocellatus*) in its stomach.

Like the great baleen whales, some of the largest sharks are adapted for filtering small organisms out of the water. Whale sharks, basking sharks and megamouth sharks feed on minute zooplankton. These sharks are equipped with wide gill slits and modified branchial structures for trapping small fish, fish eggs, crustaceans, larvae and other planktonic invertebrates. Using their branchial structures, they filter incredible quantities of water. Plankton-eater feeding techniques are similar to that used by blue sharks, which consists of swallowing whole small prey and using branchial structures to prevent the prey from escaping through the gill slits. All these sharks have small teeth. The megamouth shark has 50 rows of very small teeth on each jaw, but only three rows are functional.

Filter feeders may be immediately recognised by their large size and the extremely wide mouth, which in the whale shark and megamouth shark is terminal on the head. Foraging behaviour of giant filter-feeding sharks is

The basking shark (*Cetorhinus maximus*) actively selects areas containing a high zooplankton density, and feeds by swimming with its mouth opened wide and filtering water through its gill slits (photograph by Rohan Holt).

poorly understood owing to the problems associated with tracking them and simultaneously quantifying the food abundance. The whale shark and basking shark consume plankton by swimming at 3-5kmh with mouth opened wide, and filtering water through their wide gill slits. These giants are often found swimming just below the surface with only their first dorsal and caudal fins showing.

These sharks employ different modes of filter feeding. The feeding mode employed by the whale sharks is suction filter feeding, since they suck in a mouthful of water and purge it through their gill slits. The whale shark filtering structure is a net-like connective tissue which strains out plankton. The foraging strategy of these giants enables them to eat larger prey than the other filter feeders, including anchovies, sardines, mackerels, small tuna and squid. The whale sharks have also been seen feeding in a vertical position. They rise vertically through a school of fish, with their snout at the sea surface, then sink back with the immense mouth open wide, swallowing water and fish.

The basking shark filtering structures are long gill rakers. The feeding mode employed by the basking shark is ram-jet filter feeding, since they push the water through the gill rakers as they swim. Their huge gill slits almost encircle the head, making these fish unmistakable and enabling them to expel vast quantities of water that they have strained for food. They can filter over two thousand tons of water per hour, and it requires considerable energy to feed in this manner.

Basking sharks spend long periods feeding at the surface with their huge mouths open wide. These sharks have a huge, oily liver, and are close to neutral buoyancy. Basking shark distribution is related to zooplankton abundance. English researchers, David W Sims and Victoria A Quayle, studied basking shark foraging behaviour by tracking the movements of feeding basking sharks and by measuring the zooplankton density. Basking sharks are selective filter feeders that choose the richest plankton patches, foraging along thermal fronts, actively selecting areas with high zooplankton density, and following the productive patches. Sharks remain for up to 27 hours in rich patches that are transported by tidal currents. They move between patches over periods of one to two days, minimising travel time by following the frontal boundaries to find the closest patches. Foraging behaviour of basking sharks, therefore, indicates the distribution and density of zooplankton patches. This makes these fish unique natural plankton recorders, with potential for use as detectors of trends in abundance of zooplankton species that are influenced by climatic fluctuations.

Basking sharks are able to leap spectacularly from the sea surface. Some researchers suggest that this behaviour may remove ectoparasitic copepods and lampreys, but the reason may actually be related to feeding. Breaching near aggregations of zooplankton may cause the zooplankton to crowd closer together to form a tight mass. The tightly massed plankton is a more efficient meal for the basking shark than a dispersed group, because it offers maximum energy intake for minimum energy expenditure.

The megamouth shark, in which the mouth is terminal rather than ventral, is one of the most unusual of all sharks. This animal is equipped with gill rakers, and processes of the pharyngeal region that prevent euphasiid shrimp and other small prey from escaping.

All filter feeders have small teeth. This picture shows the teeth of a basking shark (*Cetorhinus maximus*) (photograph by Alessandro De Maddalena).

Megamouth sharks spend the daytime at a depth of 150 metres, ascend to 15 metres at night, and return to deep waters at sunrise. These movements may be a response to the movement of the plankton on which they feed. Little is known about their feeding behaviour. The megamouth shark probably feeds by swimming through aggregations of plankton, protruding its jaws and expanding its buccal cavity to suck the prey inside. It then closes its mouth, expelling water through the gill slits. The feeding mode employed by the megamouth shark is most likely suction filter feeding.

A school of fish, like these blackspotted rubberlip (*Plectorhinchus gaterinus*), is more likely than a solitary individual to detect an approaching predator and take evasive action (photograph by Vittorio Gabriotti).

SCAVENGERS

An important distinction must be made between preying on a live animal and eating a dead animal. Scavengers feed on dead organisms. Most sharks prefer live fresh food and do not like decaying meat. No species relies exclusively on scavenging to feed, but there are numerous sharks that scavenge when occasion offers. Large marine animal carcasses are known to attract sharks. Feeding on dead animals is very advantageous because it often provides enough energy to sustain a shark for long periods, with minimal energy expenditure.

Scavengers feed on a wide variety of dead animals. Fish processing plants and slaughterhouses that dump their wastes into the sea often have several sharks in attendance. Scavenging is often the reason terrestrial animals are found inside sharks (see 'Inedible items and various oddities', page 92). Fish are often torn from hooks by sharks such as tope sharks (*Galeorhinus galeus*) and requiem sharks (family Carcharhinidae). Fishery catches are important food sources for some sharks. These predators are often cautious when investigating hooked baits.

Where sharks are abundant, they can be pests to fishermen because they damage their catches. Sometimes the selachians swim under the boat and snap off each fish as it is being hauled up, leaving only the heads. In the tropical eastern Pacific Ocean, fishermen call the silky shark (*Carcharhinus falciformis*) the 'net-eater shark' because of the damage it does to nets. Recently, some members of the Indian Ocean Tuna Commission expressed their concern about the damage caused by sharks and marine mammals scavenging on longline-caught tuna. According to the Commission, losses to scavenging of 20% to 30% of the total longline catches were reported in some parts of the Indian Ocean.

Sharks that feed on hooked or netted fish are often captured while feeding. For example, thresher sharks (family Alopiidae) are often tail-hooked on pelagic longlines after striking the bait with their long caudal fin. When a large shark approaches a fishing boat and starts to eat a hooked fish, hauling the fish on board the boat can provoke an aggressive reaction, and the shark may bite the vessel's hull or in rare cases can seriously damage the boat. When large sharks swim into a gill net they frequently become trapped in the mesh and by their efforts to escape create costly damage for the fishermen. Often, the large selachians must be killed before the fishermen can retrieve their nets. Sharks caught in the protective gill nets off KwaZulu-Natal, South Africa, are often scavenged by other sharks after capture, with bites usually inflicted in the abdominal region.

Sharks such as the blacktip shark (*Carcharhinus limbatus*) and dusky shark (*Carcharhinus obscurus*) also feed on fish considered unfit for human

A great white shark (*Carcharodon carcharias*) eats a tuna used
as bait off Neptune Islands, in Spencer Gulf, South Australia.
These animals, like many other sharks, usually scavenge when
the occasion offers (photograph by Vittorio Gabriotti).

consumption which fishermen throw back into the sea. Many species follow
fishing vessels, feeding on the offal and occasionally eating the refuse from the
ship. Even speared fish are a food source that attracts sharks. The whitetip reef
shark (*Triaenodon obesus*) and the blacktip shark feed on dead or dying speared
fish, usually stealing the captured fish without attacking the diver.

Dead cetaceans, large and small, are commonly eaten by sharks. Whalers
have often observed sharks scavenging on whale carcasses, especially when the
carcasses were brought alongside the ship for the flensing. The shark feeding
behaviour on a large cetacean carcass, like that of a sperm whale (*Physeter
macrocephalus*), was described with accuracy by American writer Herman
Melville in his classic novel, *Moby Dick*. Whale carcasses may float for long

Whales are commonly eaten by sharks after death.
Each carcass provides a large amount of food for many
sharks (photograph by Marcela Moisset).

periods, forming a large slick that sends out a strong olfactory signal to sharks swimming in the area. Each carcass provides a large amount of food for many sharks. Species that commonly feed on dead whales usually have large, triangular, serrated teeth adapted for taking bites out of large carcasses.

A well-known scavenger is the great white shark (*Carcharodon carcharias*). These sharks have often been observed scavenging on carcasses of cetaceans, pinnipeds, bony fish and other sharks (including the huge basking shark, *Cetorhinus maximus*), even when these carcasses are in advanced states of decomposition. Great white sharks search for prey independently, but when a shark kills a large prey, such as a seal, the others are attracted to the site of the kill to feed. Consequently, one feeding strategy of great white sharks is to remain relatively close to one another so that, if one white shark catches prey, the others can scavenge on it. Peter Klimley and colleagues observed the appearance of other individuals shortly after a successful predation by a great white shark (they reported that in 16 out of 129 observed cases of predation, other great white sharks approached the successful predator). While there is no evidence of predation on large cetaceans, adult great white sharks depend on their carcasses as a significant portion of their diet.

Scavenged species possibly include all available cetaceans as large as huge blue whales (*Balaenoptera musculus*). Along the Atlantic coast of North America, whale carcasses are commonly accompanied by single individuals or small groups of great white sharks. The same individuals can remain around a whale carcass for at least one week. Great white sharks can also make deep dives, possibly to detect cetacean carcasses that have sunk to the sea bottom. Large cetacean carcasses can be found anywhere, and consequently they are an important food source for great white sharks worldwide.

When feeding on a floating dead cetacean, great white sharks show a general pattern when removing flesh. They sink their serrated teeth into the carcass and shake their head from side to side. The great white shark sometimes rolls on its side or rolls over on its back before biting the whale. The predator then sinks its teeth into the carcass and rotates to upright itself with powerful strokes of its wide caudal fin. This movement cuts off a large piece of meat. Great white sharks also rise above the sea surface to grab onto the carcass, using their mass and furious movements to excise a mouthful of meat. This behaviour has been described by whalemen, and was reported in a book by Peter Matthiessen called *Blue meridian*. A great white shark can remove pieces of flesh weighing about nine kilograms in a single bite while feeding. Great white sharks usually do not show upper jaw protrusion while feeding on a carcass. In order to feed on cetacean carcasses that sink to the bottom, these warm-bodied predators may also penetrate cold bottom water. Great white sharks also scavenge on other animal carcasses, but rarely on human cadavers.

Shark specialist, Geremy Cliff, and colleagues observed that 7% of the great white sharks caught in the protective gill nets off KwaZulu-Natal, South Africa, were found in the same net installation as a caught dolphin. On one occasion a great white shark was netted close to a bottlenose dolphin (*Tursiops truncatus*), and chunks of the same cetacean were found in the shark's stomach. Therefore, scavenging by great white sharks on the dolphins may be a reason for simultaneous capture. Moreover, the stomachs of these sharks often contain tuna and swordfish evidently torn from hooks.

Other sharks that often scavenge on cetaceans are the larger requiem sharks (family Carcharhinidae), such as the blue shark (*Prionace glauca*), tiger shark

(*Galeocerdo cuvier*), oceanic whitetip shark (*Carcharhinus longimanus*), bull shark (*Carcharhinus leucas*) and bronze whaler shark (*Carcharhinus brachyurus*). The species that belong to the genus *Carcharhinus* are commonly called whaler sharks because of their habit of feeding on dead and moribund whales. In Australia, the blue shark is called the blue whaler because it is commonly seen feeding on the carcasses of large cetaceans. Blue sharks can be found at whale carcasses in feeding aggregations of a dozen or more individuals. Victor G Cockcroft and colleagues studied dolphins entrapped in anti-shark nets off KwaZulu-Natal, South Africa. They observed that the dolphins that had been captured and subsequently scavenged by sharks (in particular, requiem sharks) showed characteristic signs, like the soft underbelly flesh having been removed while caudal and pectoral fins remained unconsumed. However, the most commonly implicated species was the bull shark.

The tiger shark regularly scavenges, as this large fish eats almost anything it encounters (dead or alive). Other sharks that have been commonly reported to feed on dead animals include the blacktip shark, pigeye shark (*Carcharhinus amboinensis*), grey reef shark (*Carcharhinus amblyrhynchos*), silvertip shark (*Carcharhinus albimarginatus*), sharpnose shark (*Rhizoprionodon terranovae*), great hammerhead (*Sphyrna mokarran*), bluntnose sixgill shark (*Hexanchus griseus*), broadnose sevengill shark (*Notorynchus cepedianus*), Greenland shark (*Somniosus microcephalus*) and the Pacific sleeper shark (*Somniosus pacificus*).

Many sharks will take advantage of catastrophes, caused both by humans and by nature, to feed on wounded, dying or dead animals. The scent of blood, loud sounds, and movements made by dying individuals are detected by sharks from a long distance. Catastrophes such as seaquakes, disease outbreaks, underwater explosions, shipwrecks and airplane crashes soon attract numerous sharks.

In Sicily, Italy, a terrible earthquake took place on 20 December 1908. There was a huge seaquake and enormous destruction in Messina. The large wave caused by the seaquake caused numerous deaths and disappearances. During this period, a large shark, possibly a great white shark, was found stranded in Messina: its stomach contained the leg of a woman. One month later, on 26 January 1909, another great white shark was caught off Augusta, approximately 100km south of Messina. This specimen had remains of at least three people (a man, a woman and a child) in its stomach. In both instances the victims undoubtedly were people drowned by the seaquake.

Scavenging will also be dealt with in 'Competition', page 166.

In Australia the blue shark (*Prionace glauca*) is called the blue whaler because it is commonly seen feeding on dead and moribund whales (photograph by Walter Heim).

HUNTING PREY OUT OF THE WATER

Certain sharks show unusual methods of feeding, and use behavioural techniques that enable them to catch prey along the shore or close to the sea surface. Some sharks pursue their prey to the shallowest water and rarely onto shore (a few species are able to live for a considerable time out of water). Other sharks catch sea birds, attacking them while they rest at the sea surface. The tiger shark (*Galeocerdo cuvier*), blue shark (*Prionace glauca*), great white shark

A great hammerhead shark (*Sphyrna mokarran*) ventures into the shallow waters of Walkers Cay, Bahamas. The great hammerhead shark may actually lunge on shore to grab prey (photograph by Harald Baensch).

(*Carcharodon carcharias*) and shortfin mako (*Isurus oxyrinchus*) are examples of such extraordinary selachians.

The blacktip reef shark (*Carcharhinus melanopterus*) frequently ventures into very shallow waters, and is often seen swimming with its black-tipped first dorsal fin out of the water. Blacktip reef sharks can drive fish into shallow water and onto the beach. In many cases, they swim out of the sea and onto the shore to consume the stranded fish, and then manoeuver their agile bodies back into the water (see also 'Hunting aggregated prey', page 140).

The great hammerhead shark (*Sphyrna mokarran*) may actually lunge on shore to grab prey. In one case, a great hammerhead was observed beaching itself in pursuit of a spotted eagle ray.

Blue sharks have been observed to feed only a few metres from the shore, in very shallow waters. Vinicio Biagi reported a case of several blue sharks attracted by the presence of fish remains and blood to very shallow waters close to a small tuna trap near Baratti, Italy.

Smooth-hounds (*Mustelus* sp.) feed heavily on crustaceans, and catch crabs on the muddy banks of estuaries and bays. In some cases, when crabs are at the edge of the water, the sharks check on the position of the crustaceans and then swim out to sea. Then the smooth-hounds turn and jump clear out of the water attempting to catch these crustaceans. Afterwards, they thrash around from side

s back into the sea.

hark often lifts part of its body out of the head and pectoral fins. These amazing found in other sharks, called spy-hopping, e water to investigate and locate potential ws the predator to locate pinnipeds resting nipeds another large marine predator, the chnique similar to that of the great white

white sharks lifting their bodies out of the ks. Great white sharks attracted to boats ter in order to reach baits suspended above e, even great white sharks sometimes come ar Ganzirri, in the Messina Strait, Italy, 5.6m was observed pursuing a school of from shore, where it almost touched the f its body, while its dorsal part was out of

The tiger shark is the largest shark species that commonly moves in very shallow waters to feed. Tiger sharks show a marked affinity for marine turtle

A hawksbill turtle (*Eretmochelys imbricata*) in southern
Egyptian waters of the Red Sea. These marine reptiles
are important in the diet of the tiger shark (*Galeocerdo
cuvier*) (photograph by Vittorio Gabriotti).

meat. These dangerous cartilaginous fish are found in zones where marine
turtles lay eggs. Tiger sharks venture into water barely sufficient to cover them,
and can also swim onto the beach in order to capture a turtle. A tiger shark was
observed attempting to catch a dolphin chased onto the shore. In this case, the
predator retreated back into the sea, leaving the wounded cetacean stranded on
the beach.

Tiger sharks are frequently observed feeding on sea birds. Many of the tiger
shark stomachs examined have contained sea birds, including gulls, albatross,
frigate birds, shearwaters, cormorants and pelicans. Of the 29 specimens of
tiger shark caught off New South Wales, Australia, and examined by John D
Stevens, eight contained birds. Tiger sharks are efficient specialists in capturing
albatross. At French Frigate Shoals in the north-west Hawaiian Islands, tiger
sharks annually congregate in waters where young albatross learn to fly. Tiger
sharks attack the birds when they fall to the sea during their attempts, or when
the albatross rest at the sea surface.

In these waters, researcher Wesley R Strong observed 138 attacks by tiger
sharks on young black-footed albatross (*Phoebastria nigripes*). Half of the

predatory attempts were successful. The sharks swim into the lagoon and check the position of the sea birds on the shore. The young birds make their first attempts at flying, and land on the sea surface. The sharks wait for the birds to hit the sea surface, and then attack at high speed. The faster birds have a better chance to survive than others. According to some estimates, sharks may kill nearly 10% of the young birds.

Other shark species occasionally try to capture sea birds. The shortfin mako is astonishingly athletic, and is known for its habit of leaping from the water, up to at least six metres above the sea surface, and it has been observed catching sea birds. Chris Fallows has observed makos chasing greater shearwaters, and other mid-sized pelagic birds such as subantarctic skuas, on the sea surface in South Africa; and Walter Heim has witnessed makos trying to capture shearwaters

The shortfin mako (*Isurus oxyrinchus*) occasionally tries to capture sea birds (photograph by Walter Heim).

A 3.25m shortfin mako (*Isurus oxyrinchus*) caught off Port
Mansfield, Texas, USA, on 27 January 2002. Its stomach
contained ten sea birds, probably gulls (photograph by
Jeff Shindle).

off San Diego, California, USA. In all these cases, the birds were able to escape the predators, but Heim has noticed many gulls and other birds with missing or mauled legs, probably due to a shark attack; and we know that some of these predatory attempts are successful. The stomach of a 3.25m shortfin mako caught off Port Mansfield, Texas, USA, contained ten sea birds, probably gulls.

Even the great white shark has been observed pursuing sea birds, such as African penguins (*Spheniscus demersus*) and gulls, resting at the sea surface. A great white shark trying to capture mutton birds resting on the water surface in South Australian waters has been described by Wesley Rocky Strong Jr. An interesting observation concerning shark predation on birds was made by Harry Blake-Knox: in Dublin he observed a common thresher shark (*Alopias vulpinus*) killing a wounded loon, a diving bird (*Gavia* sp.), with a slap of the incredibly long upper lobe of its caudal fin, and then eating it.

COOPERATIVE HUNTERS

In general, sharks occur singly, in pairs, or in groups of varying size depending on species. There are also some sharks that are rarely seen in the company of other individuals of the same species.

Most sharks are not highly social animals. They do not live in groups with highly organised and complex social systems. It is not surprising that most sharks hunt alone. Simply, most selachians do not depend on one another for catching and killing prey. However, although solitary behaviour has been observed for many species, other species do catch in groups.

Are sharks solitary or cooperative hunters? Cooperative hunters are two or more sharks, of the same or different species, actually collaborating to detect and kill prey. These relationships should be mutualistic, where all participants benefit from the relationship by capturing prey more easily or by feeding upon larger prey. We know that sharks are capable of sophisticated communication, but the problem is that we do not know if and how individuals of a given species that hunt in a group actually interact to locate and capture prey. For example, some accounts suggest that one shark attacks a prey animal from one side, while another shark executes a feeding assault from the opposite direction. These predatory attacks are apparently coordinated, but it is difficult to determine whether these simultaneous attacks are actually made with the intention of cooperating or are simply a matter of chance. A few researchers have tried to observe the interactions among sharks hunting, but much about these behaviours is still unknown. Several shark species most likely hunt their prey cooperatively, although most observations are anecdotal in nature. For a part of their lives, some species probably hunt in groups.

A shark species that has a very interesting social behaviour is the smooth hammerhead (*Sphyrna zygaena*). These sharks move in spectacular schools in many parts of the world, including South African waters, the eastern coast of the United States, and the Messina Strait in Italy. Usually these congregations, often hundreds of individuals, consist of young individuals measuring up to 1.8m. Italian researcher Antonio Celona reported that groups of smooth hammerheads, usually of eight to 12 individuals (probably the visible part of larger groups), passed through the Messina Strait, Sicily, following the frigate mackerels (*Auxis thazard*) on which they were observed feeding. Strangely, immense schools formed by the closely related scalloped hammerhead (*Sphyrna lewini*) do not seem to be related to feeding. These sharks do not show any interest in food during daylight hours when they congregate, but they feed at night when the

The blue shark (*Prionace glauca*) catches prey in groups,
but it is unknown if and how individuals actually interact
to locate and capture prey (photograph by Walter Heim).

school disperses. The function of these gatherings is generally unknown, but
may be related to mating and protection of the young.

Some sharks, including the sandtiger shark (*Carcharias taurus*), grey reef
shark (*Carcharhinus amblyrhynchos*) and blacktip reef shark (*Carcharhinus
melanopterus*), form aggregations in order to drive and herd schooling fish
into shallow waters, where they attack and kill their prey (see also 'Hunting
aggregated prey', page 140). These sharks appear to interact in the capture of
prey. Sandtiger sharks can occasionally form enormous schools of up to two
hundred individuals. Russell J Coles eyewitnessed and described the following
sandtiger shark predatory tactic: a school of a hundred sharks surrounded a
school of bluefish (*Pomatomus saltatrix*) and forced them into a tightly packed
school in shallow water. Then the entire school of sandtiger sharks simultaneously
attacked them.

The stomachs of lanternsharks (genus *Etmopterus*) can contain parts of large squid. Lanternsharks may be schooling predators, and may use their photophores to coordinate their school as they attack large squid or schools of small fish. Salmon sharks (*Lamna ditropis*) are often encountered in groups feeding on Pacific salmon (*Oncorhynchus* sp.). Groups of common thresher sharks (*Alopias vulpinus*) sometimes attack schools of sardines. Even broadnose sevengill sharks (*Notorynchus cepedianus*) form groups to catch sea lions. Other species that catch in groups include the tope shark (*Galeorhinus galeus*), whitetip reef shark (*Triaenodon obesus*), spiny dogfish (*Squalus acanthias*), longnose spurdog (*Squalus blainvillei*) and the basking shark (*Cetorhinus maximus*). However, we do not know if and how individuals of these species actually interact to locate and capture prey.

The great white shark (*Carcharodon carcharias*) appears to be a solitary hunter. During a study at Año Nuevo Island, California, USA, researchers found little evidence of social cooperation in great white sharks while they were hunting the sea lion and seal colony. Rarely did a shark actually approach another individual, and they were rarely less than a hundred metres from each other. The sharks often arrived and departed simultaneously, but at other times arrived and departed separately. Generally, they did not swim together while near the island. The great white sharks were not hunting as a social group, but searched for prey independently. Simply, when a shark attacked a pinniped, the others were attracted to the site of the kill to feed (see also 'Scavengers', page 150). While juvenile white sharks might school together, adults are solitary. The age at which these sharks stop schooling and become solitary individuals is unknown.

A sandtiger shark (*Carcharias taurus*). A group of a hundred sandtiger sharks has been observed surrounding a school of bluefish (*Pomatomus saltatrix*) and forcing them into a tightly packed school in shallow water before simultaneously attacking them (photograph by Christopher Parsons / Tennessee Aquarium).

COMPETITION

Competition occurs when two or more sharks simultaneously attempt to feed on the same prey. Researchers who have studied shark behaviour have observed both interspecific and intraspecific competition.

Access to food is often established through agonistic behaviour, in which individuals of some species are very aggressive towards one another. Evidence indicates the existence of individual combat among sharks. Wounds that are the result of intraspecific aggression have been observed on the bodies of some shark species, in both males and females, and in both mature and immature individuals. These wounds are often more severe than 'love bites'. This kind

Many sharks bear scars and scrapes from fights with conspecifics (photograph by Vittorio Gabriotti).

of wound has been observed, for example, on some shortfin makos (*Isurus oxyrinchus*) and great white sharks (*Carcharodon carcharias*). Cannibalism is relatively common in sharks, and many sharks feed on other shark species.

In rare cases, agonistic behaviour takes the form of real fighting. In fact, most scars resulting from intraspecific combat do not last more than a few years. These agonistic behaviours have a communication function, as the predator attempts to communicate with conspecifics and individuals of other species using particular signals before attacking. These behaviours may function as a warning, telling other sharks that it is ready to attack and able to inflict severe injury. Combats can be bloody, and one of the adversaries may die as a result of the wounds suffered during these ferocious battles.

Agonistic behaviour can prevent direct fights, forcing the other individual to flee without resorting to fighting. The blacknose shark (*Carcharhinus acronotus*) and the bonnethead shark (*Sphyrna tiburo*) respond to the presence of conspecifics with a threat display in which they swim with an arched back, snout lifted and caudal fin lowered. A similar threat display is performed by grey reef sharks (*Carcharhinus amblyrhynchos*) and Galapagos sharks (*Carcharhinus galapagensis*) when disturbed by divers, but may also be directed towards conspecifics (see 'Shark attacks on humans', page 82).

Most sharks are not highly social animals, but they often form temporary social structures. Social hierarchies between different species and among members of the same species have been reported when sharks are feeding. Hierarchies serve as an anti-predatory tactic on the part of the subordinate shark.

An order of dominance has been shown to exist based on species, size and gender. Great white sharks dominate blue sharks (*Prionace glauca*) when both species are feeding. According to American ichthyologist Douglas J Long, blue sharks do not scavenge on a whale carcass when white sharks are feeding. Oceanic whitetip sharks (*Carcharhinus longimanus*) are dominant over silky sharks (*Carcharhinus falciformis*) of similar size. Silvertip sharks (*Carcharhinus albimarginatus*) dominate Galapagos sharks, while both are dominant over blacktip sharks (*Carcharhinus limbatus*). Off San Diego, California, photographer Richard Herrmann has observed that when a larger blue shark (*Prionace glauca*) approaches a smaller shortfin mako (*Isurus oxyrinchus*), the mako often leaves the area. Herrmann also eyewitnessed the rare event of a 2.5m blue shark attack and eat a 1m mako. In the same waters, Walter Heim has often observed 1.2-1.5m makos chase away smaller 0.6-1.2m blues from around the chum bucket.

Social hierarchies among members of the same species have been reported, for example, in shortfin mako sharks, great white sharks (*Carcharodon carcharias*), bonnetheads (*Sphyrna tiburo*) and silvertip sharks. These hierarchies are based on size. Sharks must possess a keen awareness of their own size. Consequently, smaller sharks usually move away from larger members of their own species. In order to compare their relative size or to intimidate a conspecific, pairs of great white sharks have been observed swimming parallel to each other, until one of the two surrenders to the dominant individual and accelerates away. In other cases, a pair of great white sharks swim on a collision course and the subordinate individual gives way.

Competition is evident between sharks attracted near a food source. Off San Diego, California, photographer Walter Heim has observed that shortfin makos show a hierarchy around the chum bucket, with the larger makos displacing the smaller ones. They will sometimes swim toward each other, turning at the last second and leaving a big splash. You can often tell when another shark is around by observing the behaviour of the sharks around the chum bucket, as they seem nervous.

Small makos will often show in pairs. The pairs are usually individuals of the same size, and tend to tolerate each other. Unlike makos, larger blue sharks seem to tolerate smaller sharks. Avoidance behaviours have been noticed during feeding on large dead cetaceans. During the filming of the famous documentary, 'Blue Water, White Death', off Durban, South Africa, there were over twenty oceanic whitetip sharks circling the carcass of a sperm whale (*Physeter macrocephalus*), but writer Peter Matthiessen reported that only two or three individuals fed on the cetacean at the same time. The same observation was also made by photographer Stan Waterman, when over a hundred requiem sharks (family Carcharhinidae) swam around the carcass of another sperm whale.

When a number of great white sharks are present around a carcass, only one or two individuals feed at a time. They seem to take turns feeding, and each individual feeds for about the same amount of time. In some cases, when two sharks feed simultaneously, they stay on opposite sides of the carcass. Douglas J Long observed five great white sharks scavenging on a blue whale (*Balaenoptera musculus*) carcass off San Francisco, California, USA, but only one shark fed at a time. Off Long Island, New York, USA, when at least nine great white sharks were observed visiting a fin whale (*Balaenoptera physalus*) carcass, only one or two individuals fed at the same time. A great white shark of an estimated 3-4m was seen approaching the dead whale, but suddenly left without feeding, and an instant later a 5-6m great white shark approached the same part of the carcass and began to consume it. Many of the great white sharks seen near the whale had lacerations that were probably inflicted by members of their own species while competing for the carcass.

Tobey Curtis, Karl Menard and Karl Laroche observed three great white sharks scavenging on a humpback whale (*Megaptera novaeangliae*) carcass at

168

Wounds that are the result of intraspecific aggression have been observed on the bodies of some shark species, in both males and females, and in both mature and immature individuals. These wounds are often more severe than 'love bites' (photograph by Walter Heim).

Point Reyes National Seashore, California, USA. Researchers observed no strong competition among these sharks, but only one or two individuals fed at the same time, and when two sharks fed simultaneously they were on opposite sides of the carcass. The great white sharks seemed to take turns feeding.

Competition is even more evident among sharks attracted near a smaller animal carcass. In this situation sharks often contest ownership of the food item. When a great white shark kills prey such as a marine mammal, other great white sharks are attracted to the site of the kill to feed. In these instances, competitors have to be dissuaded by prompt aggressive reaction. The first shark responds to these uninvited guests with threat displays indicative of competition for the prey. A 'tail slap' involves a pair of sharks that lift the caudal fin and splash water at each other. An individual is permitted to eat the prey only if the vigour and frequency of its tail slap is greater than that of its competitor. A 'breach' is when a shark leaps out of the water, two-thirds of the body emerging at an angle of 30°-60° to the sea surface, and is a less common behaviour that may be a higher-intensity display.

The great white shark can also perform other threat displays, such as gaping its lower jaw slightly or even protruding the upper jaw, or lowering both pectoral fins. Another interesting great white shark behaviour called 'repetitive aerial gaping' has been described by Wesley Rocky Strong Jr. When a great white shark is prevented from reaching a bait when the wrangler pulls it away, the predator tries to seize it by holding its head out of the water and rolling onto its side. It then opens and closes its mouth slowly, displaying partial gapes while swimming slowly at the surface. The repetitive aerial gaping is not oriented toward the food or other objects. Frustration can give rise to aggression, and under similar conditions great white sharks have been observed to bite conspecifics. Wesley Rocky Strong Jr has hypothesised that repetitive aerial gaping may be a manifestation of frustration, and may function to reduce intraspecific aggression, redirecting frustration and avoiding attacks on other great white sharks.

Are sharks territorial animals? Do they defend a given area as their territory? In general, it seems most sharks do not possess any territory. They often show philopatry, or a special preference for an area where they stay or to which they return periodically, but they do share this territory with other members of their species and other sharks, and appear to coexist without much conflict. In a study on great white sharks conducted in the waters surrounding pinniped colonies at Año Nuevo Island, California, USA, researchers observed an almost complete absence of territoriality. Although each of the great white sharks spent more time in a slightly different location than the other individuals, and some sharks patrolled certain areas preferentially, all the individuals frequently moved over the same areas. There was no evidence that each individual defended an area as territory.

Blue sharks (*Prionace glauca*). Sharks often segregate by size: this behaviour reduces the risk of intraspecific aggression (photograph by Walter Heim).

An important part of the global food chain

Sharks are apex predators, and we can imagine them at the top of an imaginary pyramid, called the pyramid of biomass. This pyramid represents the total amount of energy and living mass in an ecosystem. The greatest amount of energy and biomass is present at the base of the pyramid with the producers (plants), while the least amount of energy and biomass is found at the top of the pyramid with the highest level of consumers or apex predators. Energy decreases steeply, and so does the biomass that can be supported at each level. The smallest organisms are very numerous, and supply nourishment to the next largest. As we move up this pyramid, organisms increase in size and decrease in quantity.

Sharks are at the top of most marine food chains, and they have virtually no enemies. Their young and eggs are at risk of being prey to a restricted range of animals, like some fish and molluscs. Only a few creatures prey on adult sharks, such as other sharks, humans, a few bony fish like large groupers, and a few marine mammals including the killer whale (*Orcinus orca*), sperm whale (*Physeter macrocephalus*) and California sea lion (*Zalophus californianus*). The size of most sharks is sufficiently large to preclude them from predation by other animals.

However, sharks are hosts to numerous parasites, such as copepods, isopods, lice, leeches, trematods, cestodes, and nematodes. These parasite organisms live in and on the cartilaginous fish and obtain sustenance from them. Parasites can cause serious lesions, sometimes resulting in severe diseases. Amount and species of parasites depend on the species of shark; for example, off California, Walter Heim observed that shortfin mako sharks (*Isurus oxyrinchus*) are often encrusted with copepod parasites, even at an early age, while blue sharks (*Prionace glauca*) typically have few or no copepod parasites.

Sharks play an important ecological role in marine communities. These fish, as predators, are fundamental instruments of natural selection. Moreover, as scavengers they help process organic material, so that it can then be used by other animals and plants. They influence the composition of marine ecosystems, contribute to their stability, and maintain biodiversity.

Sharks have a substantial impact on prey organisms, as shark predation is an important natural control on population size of many marine species. These fish play a very significant role in marine food chains. For example, off the north-east coast of the USA, the primary food source of the shortfin mako consists of

Sharks always have parasites, such as copepods, small crustaceans only a few millimetres or centimetres long. This shortfin mako (*Isurus oxyrinchus*) was found with copepods attached to the mouth and first dorsal fin (photograph by Walter Heim).

bluefish (*Pomatomus saltatrix*). According to fisheries biologist Chuck Stillwell, the mass of bluefish eaten by shortfin makos in the area from Cape Hatteras, North Carolina, to Georges Bank, New England, is estimated to be 7.3% of the bluefish in this region! Pacific salmons (*Oncorhynchus* spp.) are the major prey item of salmon sharks (*Lamna ditropis*) in subarctic waters. According to fisheries biologist Kazuya Nagasawa, from spring to autumn 1989 salmon sharks five years old or older occurring in subarctic waters appear to have consumed 12.6–25.2% of the total annual run of Pacific salmon. Knowing the food requirements of sharks and their abundance is fundamental in understanding the effect these predators have on marine ecosystems.

Shortfin makos (*Isurus oxyrinchus*) at the Milano fish market, Italy. Many sharks are fished commercially in order to obtain meat and other products (photograph by Alessandro De Maddalena).

Sharks have much more cause to fear humans than vice versa. Human beings are the only predators affecting shark survival, as many species are of economic importance. In 1968 shark specialist Perry W Gilbert suggested that shark fishery and commercial utilisation be increased. Today the situation has changed, and sharks are decreasing in all the oceans as a result of human activities. Many sharks are fished commercially, and are actually overfished in many seas of the world. Many species, such as the porbeagle (*Lamna nasus*), shortfin mako, piked dogfish (*Squalus acanthias*), smooth-hound (*Mustelus* sp.) and requiem shark (family Carcharhinidae) are heavily exploited. The piked dogfish is the leading commercial shark taken in the world.

An estimated 50% of the world shark catch is believed to be taken accidentally while fishing for other species such as tuna and swordfish. This unplanned capture of marine animals is called 'bycatch'. Pelagic longlines are single-stranded fishing lines 18 to 72km long, with an average of 1 500 baited hooks. This gear is widely used in many parts of the world to catch tuna and swordfish. In some areas, the number of sharks caught by longliners reaches 90% of total captures. Species such as the blue shark and shortfin mako are strongly affected by this fishing strategy.

As bony fish fisheries have been depleted, fishermen have compensated by increasing shark captures. Fishermen are reducing numerous shark populations more rapidly than those of most bony fish. In fact, sharks are more vulnerable to overfishing than are bony fish. Since few species prey on them, sharks are highly vulnerable to over-exploitation, especially since they have long sexual maturation times, low fecundity and long gestation periods, and produce small numbers of young. Moreover, many shark species segregate by size and sex, such that exploitation of sharks in a nursery area can be particularly devastating. These fish are unable to withstand extended periods of over-exploitation, which has long-term effects, and rebuilding shark populations takes many years. Most commercial shark fisheries collapse within a few years.

Humans catch sharks in order to obtain meat, cartilage, skin, oil and other products. In many countries shark meat is a significant part of the human diet. Shark fins are used in Chinese cooking to prepare a famous shark fin soup. Recently, the demand for shark fins has increased dramatically. Shark fins are high-priced, and this has led to the practice of finning sharks at sea, where the fins are sliced off while the rest of the body is discarded overboard. Almost all large and medium-sized sharks are fished for their fins. The flesh also makes a good supplement for animal feed. Shark cartilage is used in pharmaceuticals. The liver is rich in vitamins and provides oil and squalene, which are used for lubricants, cosmetics and pharmaceuticals. The skin is used to obtain a particular leather and shagreen. Shark corneas are used as substitutes for human corneas. Teeth, jaws and taxidermied specimens are used for decoration and as souvenirs.

Many species of shark have become uncommon or rare owing to overfishing of either the shark or its prey. Nobody knows how many sharks are caught in the world, but the number is estimated to be enormous. Annual landings of cartilaginous fish reported to the Food and Agriculture Organization (FAO) of the United Nations amount to around 800 000 tons, but the actual total is surely much higher since great amounts of catch are not recorded. This estimate does not include several thousands of sharks that are killed by fishermen annually and thrown back into the sea because in many countries numerous shark species are considered non-marketable.

Industrial fishing vessels often operate in flagrant violation of fishing regulations. Moreover, many species are caught by recreational anglers. Humans also have a less direct, but just as harmful, effect on sharks because of depletion of resources, environmental pollution and habitat destruction. Toxic chemicals that can be absorbed or ingested by animals are passed up the food chain through consumption. Consequently, top predators like sharks are at higher risk since several toxins accumulate in each organism along the food chain, becoming most concentrated at the top.

A few species are now protected in some countries, but it is not enough. The decline in sharks warrants an urgent investigation into the status of the species. Effective conservation and management of shark fisheries are based on research in biology, ecology, distribution, abundance and exploitation of sharks and their prey. Understanding the biology and ecology of sharks can provide significant insights into protecting them from extinction.

Important advancement in our knowledge of sharks is needed, as much about the biology and behaviour of many sharks is still not well known. Despite sharks being important parts of marine ecosystems, shark research is often neglected in favour of research into the more commercially important bony fish. The need for biological information on the life history of many shark species, including feeding ecology and predator-prey relationships, is necessary in order better to assess stock status and harvest impact. In spite of their size, surprisingly little is known about the feeding biology of many species.

It is also necessary more effectively to manage fisheries in which sharks constitute a significant bycatch. Lack of research and management in many countries is leading to the extinction of many shark species.

The removal of sharks upsets the ecological balance, which will lead to increases in some prey populations and consequently to declines in other prey species. Habitat health is dependent on all the animals that share it. The stability of marine ecosystems is in serious danger, and a worldwide effort is needed to maintain ocean wildlife heritage.

Shark fins drying in Taiwan. Fins are used
in Chinese cooking to prepare shark fin
soup (photograph by R Chen / WildAid).

APPENDIX I:
HOW TO CONTRIBUTE TO THE
STUDY OF SHARK FEEDING BIOLOGY

Researchers studying diet and predatory tactics of sharks are extremely interested in any observation of the feeding biology of these animals. Therefore any detailed account and any supporting documentation of the shark and its prey is helpful. All shark species are interesting, not only large species.

Anybody can participate in the research. If you are interested in shark feeding biology, it is advisable to maintain contacts with the fishermen working in your area and try to examine the fish they catch.

To help in data collection, a simple form has been prepared, and is included in this section. If you would like to contribute to on-going research, please complete the form and send it to the address indicated. The form includes basic data only, so that it applies to any situation. Fill in the form as accurately as possible. Always try to enclose a photo of the specimen, but if that is not possible indicate the characteristics on which the species identification is based. The shark total length should be measured in a straight line, from the snout to the tip of the upper lobe of the caudal fin. Specify if the reported measurement is only an estimate. The shark weight should be taken on the whole specimen; specify if the specimen has been gutted. Also specify if the reported weight is only an estimate. Remember that the sex of the shark is easily recognisable by observing the underside of the animal, where males have claspers (copulatory organs), which are two cylindrical appendages developed from the pelvic fin bases (note that claspers are less developed in young).

Take a photo of the whole shark, from the side. Also, take photos of the shark stomach contents, shark teeth, shark bite scars and any fresh wounds found on a victim, and any tooth enamel fragments removed from the wounds of a victim. Use a normal or short telephoto lens (not a wide-angle lens) and include a scale bar (metre stick or equivalent) in the photo. Please also specify whether or not you authorise the publication of your data and photos. Any additional documentation (photos, videos or other) should be sent together with the completed form. You can send the material via postal mail or via electronic mail.

Your help will be very much appreciated, and the author thanks in advance all who will assist in this initiative.

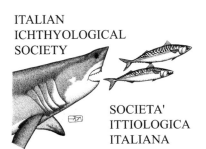

ITALIAN
ICHTHYOLOGICAL
SOCIETY

SOCIETA'
ITTIOLOGICA
ITALIANA

FORM FOR REPORTING OBSERVATIONS OF SHARK FEEDING BIOLOGY

Kind of observation:

a) examination of shark stomach contents through dissection of a dead specimen;

b) observation of shark predation or scavenging;

c) examination of signs of shark predation or scavenging on a carcass or a wounded animal:

Shark species[1]: ....

Number of sharks observed: ....

Date of shark capture or predatory event: ....

Time of day of shark capture or predatory event: ....

Location of shark capture or predatory event: ....

Distance from shore: ....

Sea depth: ....

Shark total length[2]: ....

Shark weight[3]: ....

Shark sex[4]: ....

Shark stomach contents: ....

Prey species: ....

Number of prey observed: ....

Prey total length[5]: ....

Describe in detail the event you observed[6]: ....

<u>Always enclose any supporting documentation of the shark and its prey (photos, videos or other).</u>

DATA OF THE COMPILER

Name:
....

Address:
....

....

....

Telephone:
....

E-mail:
....

Please specify whether or not you authorise the publication of your data and photos:

NOTES:

(1) If enclosing a photo of the shark is not possible, indicate the characteristics on which the species identification is based.

(2) In a straight line, from the snout to the tip of the upper lobe of the caudal fin.

(3) Specify if whole or gutted.

(4) On their underside males have two cylindrical appendages at the pelvic fin base.

(5) In a straight line.

(6) Attach an extra page if necessary.

Please send this form (via postal mail or via electronic mail) to the following address:

**Dr Alessandro De Maddalena – Italian Ichthyological Society
via L Ariosto 4, I-20145 Milan, Italy
E-mail: a-demaddalena@tiscali.it**

APPENDIX II:
CLASSIFICATION OF SHARKS
CITED IN THE TEXT

ORDER HEXANCHIFORMES

Family Chlamydoselachidae
Chlamydoselachus anguineus – Frilled shark

Family Hexanchidae
Heptranchias perlo – Sharpnose sevengill shark
Hexanchus griseus – Bluntnose sixgill shark
Notorynchus cepedianus – Broadnose sevengill shark

ORDER SQUALIFORMES

Family Squalidae
Centrophorus granulosus – Gulper shark
Centrophorus squamosus – Leafscale gulper shark
Centroscyllium fabricii – Black dogfish
Centroscymnus coelolepis – Portuguese shark
Cirrhigaleus barbifer – Mandarin dogfish
Dalatias licha – Kitefin shark
Etmopterus spinax – Velvet belly
Euprotomicrus bispinatus – Pygmy shark
Isistius brasiliensis – Cookiecutter shark
Isistius plutodus – Largetooth cookiecutter shark
Somniosus microcephalus – Greenland shark
Somniosus pacificus – Pacific sleeper shark
Squalus acanthias – Piked dogfish or spiny dogfish
Squalus blainvillei – Longnose spurdog
Squalus megalops – Shortnose spurdog

ORDER SQUATINIFORMES

Family Squatinidae
Squatina californica – Pacific angelshark
Squatina oculata – Smoothback angelshark

ORDER HETERODONTIFORMES

Family Heterodontidae
Heterodontus francisci – Horn shark
Heterodontus galeatus – Crested bullhead shark
Heterodontus portusjacksoni – Port Jackson shark

ORDER ORECTOLOBIFORMES

Family Brachaeluridae
Brachaelurus waddi – Blind shark

Family Orectolobidae
Orectolobus maculats – Spotted wobbegong

Family Hemiscyllidae
Hemiscyllium ocellatum – Epaulette shark

Family Stegostomatidae
Stegostoma fasciatum – Zebra shark

Family Ginglymostomatidae
Ginglymostoma cirratum – Nurse shark
Nebrius ferrugineus – Tawny nurse shark

Family Rhiniodontidae
Rhincodon typus – Whale shark

ORDER LAMNIFORMES

Family Odontaspididae
Carcharias taurus – Sandtiger shark

Family Mitsukurinidae

Mitsukurina owstoni – Goblin shark

Family Megachasmidae

Megachasma pelagios – Megamouth shark

Family Alopiidae

Alopias superciliosus – Bigeye thresher
Alopias vulpinus – Common thresher shark

Family Cetorhinidae

Cetorhinus maximus – Basking shark

Family Lamnidae

Carcharodon carcharias – Great white shark
Isurus oxyrinchus – Shortfin mako
Isurus paucus – Longfin mako
Lamna ditropis – Salmon shark
Lamna nasus – Porbeagle

ORDER CARCHARHINIFORMES

Family Scyliorhinidae

Apristurus brunneus – Brown catshark
Apristurus herklotsi – Longfin catshark
Apristurus longicephalus – Longhead catshark
Apristurus microps – Smalleye catshark
Cephaloscyllium isabellum – Draughtsboard shark
Cephaloscyllium ventriosum – Swellshark
Parmaturus xaniurus – Filetail catshark
Poroderma africanum Striped catshark
Poroderma marleyi – Barbeled catshark
Poroderma pantherinum – Leopard catshark
Schroederichthys maculatus – Narrowtail catshark
Scyliorhinus canicula – Small-spotted catshark

Family Triakidae
Galeorhinus galeus – Tope shark or soupfin shark
Mustelus asterias – Starry smooth-hound
Mustelus canis – Dusky smooth-hound
Mustelus henlei – Brown smooth-hound
Mustelus mustelus – Smooth-hound
Mustelus mediterraneus – Blackspotted smooth-hound
Triakis semifasciata – Leopard shark

Family Hemigaleidae
Hemigaleus microstoma – Sicklefin weasel shark

Family Carcharhinidae
Carcharhinus acronotus – Blacknose shark
Carcharhinus albimarginatus – Silvertip shark
Carcharhinus amblyrhynchos – Grey reef shark
Carcharhinus amboinensis – Pigeye shark
Carcharhinus brachyurus – Bronze whaler shark or copper shark
Carcharhinus brevipinna – Spinner shark
Carcharhinus falciformis – Silky shark
Carcharhinus galapagensis – Galapagos shark
Carcharhinus leucas – Bull shark
Carcharhinus limbatus – Blacktip shark
Carcharhinus longimanus – Oceanic whitetip shark
Carcharhinus melanopterus – Blacktip reef shark
Carcharhinus obscurus – Dusky shark
Carcharhinus perezi – Caribbean reef shark
Carcharhinus plumbeus – Sandbar shark
Galeocerdo cuvier – Tiger shark
Negaprion brevirostris – Lemon shark
Prionace glauca – Blue shark
Rhizoprionodon acutus – Milkshark
Rhizoprionodon terranovae – Sharpnose shark
Triaenodon obesus – Whitetip reef shark

Family Sphyrnidae
Sphyrna lewini – Scalloped hammerhead
Sphyrna mokarran – Great hammerhead
Sphyrna tiburo – Bonnethead
Sphyrna zygaena – Smooth hammerhead

BIBLIOGRAPHY

AINLEY, DG, STRONG, CS, HUBER, HR, LEWIS, TJ and SH MORRELL (1981): Predation by sharks on pinnipeds at the Farallon Islands. *Fishery Bulletin*, 78: 941-945.

AITKEN, K (1998): *Sharks and rays of Australia*. New Holland Publishers, Australia, 96 pp.

AMES, JA, GEIBEL, JG, WENDELL, FE and CA PATTISON (1996): White shark-inflicted wounds of sea otters in California, 1968 1992. Pp. 309-316 in Klimley, AP and DG Ainley (eds): *Great white sharks: The biology of* Carcharodon carcharias. Academic Press, San Diego, 518 pp.

AMES, JA and GV MOREJOHN (1980): Evidence of white shark, *Carcharodon carcharias*, attacks on sea otters, *Enhydra lutris*. *California Fish and Game*, 66 (4): 196-209.

ANDERSON, SD, HENDERSON, RP, PYLE, P and DG AINLEY (1996): White shark reactions to unbaited decoys. Pp. 223-228 in Klimley, AP and DG Ainley (eds): *Great white sharks: The biology of* Carcharodon carcharias. Academic Press, San Diego, 518 pp.

ANDERSON, SD, KLIMLEY, AP, PYLE, P and RP HENDERSON (1996): Tidal height and white shark predation at the Farallon Islands, California. Pp. 275-279 in Klimley, AP and DG Ainley (eds): *Great white sharks: The biology of* Carcharodon carcharias. Academic Press, San Diego, 518 pp.

ANDERSON, SS: Recurrent California sea lion predation upon a Southern California leopard shark population. Unpublished manuscript.

ARNOLD, PW (1972). Predation on the harbour porpoise, *Phocoena phocoena*, by a white shark, *Carcharodon carcharias*. *Journal Fisheries Research Board of Canada*, 29 (8): 1213-1214.

BARRULL, J (1994): La dentición como parámetro de estudio en la alimentación de los tiburones. *Quercus*, 105; 19-22.

BARRULL, J and I MATE (2002): *Tiburones del Mediterráneo*. Llibreria El Set-ciències, Arenys de Mar, 292 pp.

BENZ, GW, BORUCINSKA, JD, LOWRY, LF and HE WHITELEY (2002): Ocular lesions associated with attachment of the copepod *Ommatokoita elongata* (Lernaeopodidae: Siphonostomatoida) to corneas of Pacific sleeper sharks *Somniosus pacificus* captured off Alaska in Prince William Sound. *The Journal of Parasitology*, 88 (3): 474-481.

BENZ, GW, LUCAS, Z and LF LOWRY (1998): New Host and Ocean Records for the Copepod *Ommatokoita elongata* (Siphonostomatoida: Lernaeopodidae), a Parasite of the Eyes of Sleeper Sharks. *The Journal of Parasitology*, 84 (6): 1271-1274.

BIAGI, V (1995): *Memorie della 'Tonnara' di Baratti*. 2ª edizione. 1835-1939. Circolo Nautico Pesca Sportiva Baratti, Venturina, 96 pp.

BIGELOW, HB and WC SCHROEDER (1948): *Sharks*. Pp. 53-576 in *Fishes of the Western North Atlantic Part one: Lancelets, Ciclostomes, Sharks*. Memoir Sears Foundation for Marine Research. Yale University, New Haven, 576 pp.

BORUCINSKA, JD, BENZ, GW and HE WHITELEY (1998): Ocular lesions associated with attachment of the parasitic copepod *Ommatokoita elongata* (Grant) to corneas of Greenland sharks, *Somniosus microcephalus* (Bloch and Schneider). *Journal of Fish Diseases*, 21 (6): 415.

BUENCUERPO, V, RIOS, S and J MORON (1998): Pelagic sharks associated with the swordfish, *Xiphias gladius*, fishery in the eastern North Atlantic Ocean and the Strait of Gibraltar. *Fishery Bulletin*, 96 (4): 667-685.

BURGESS, GH (1991): Shark attack and the International Shark Attack File. Pp. 101-105 in Gruber, SH (ed): Discovering sharks. *Underwater Naturalist, Bulletin American Littoral Society*, 19 (4)-20 (1).

BURGESS, GH and M CALLAHAN (1996): Worldwide patterns of white shark attacks on humans. Pp. 457-469 in Klimley, AP and DG Ainley (eds): *Great white sharks: The biology of* Carcharodon carcharias. Academic Press, San Diego, 518 pp.

CADENAT, J and J BLACHE (1981): Requins de Méditerranée et d'Atlantique (plus particulièrement de la Côte Occidentale d'Afrique). *Faune Tropicale, ORSTOM*, Paris, 21: 1-330.

CAPAPE', C (1989): Les Sélaciens des côtes méditerranéennes: aspects generaux de leur écologie et exemples de peuplements. *Océanis*, 15 (3): 309-331.

CAREY, FG, KANWISHER, JW, BRAZIER, O, GABRIELSON, G, CASEY, JG and HL PRATT (1982): Temperature and activities of a white shark, *Carcharodon carcharias*. *Copeia*, 1982: 254-260.

CASEY, JG and HL PRATT (1986): White sharks in the Western North Atlantic. *Maritimes*, November 1986: 4-6.

CASTRO, J (1983): *The Sharks of North American Waters*. Texas A&M University Press, College Station, 180 pp.

CELONA, A, DE MADDALENA, A and T ROMEO (2005): Bluntnose sixgill shark, *Hexanchus griseus* (Bonnaterre, 1788), in the eastern north Sicilian waters. *Bollettino del Museo Civico di Storia Naturale di Venezia*, 56: 137-151.

CELONA, A, DE MADDALENA, A and G COMPARETTO (2006): Evidence of a predatory attack on a bottlenose dolphin *Tursiops truncatus* by a great white shark *Carcharodon carcharias* in the Mediterranean Sea. *Annales, Series historia naturalis*, 16 (2): 159-164.

CELONA, A, DONATO, N and A DE MADDALENA (2001): In relation to the captures of a great white shark, *Carcharodon carcharias* (Linnaeus, 1758), and a shortfin mako, *Isurus oxyrinchus* (Rafinesque, 1809), in the Messina Strait. *Annales, Series historia naturalis*, 11 (1): 13-16.

CHAPMAN, D and SH GRUBER (2002): A further observation of the prey-handling behaviour of the great hammerhead shark, *Sphyrna mokarran*: predation on the spotted eagle ray, *Aetobatus narinari*. *Bulletin of Marine Science*, 70 (3): 947-952.

CHE-TSUNG, C, KWANG-MING, L and J SHOOU-JENG (1997): Preliminary Report on Taiwan's Whale Shark Fishery. *TRAFFIC Bulletin*, 17 (1).

CLARK, E and E KRISTOF (1991): How deep do sharks go? Reflections on deep sea sharks. Pp. 77-78 in Gruber, SH (ed): Discovering sharks. *Underwater Naturalist, Bulletin American Littoral Society*, 19 (4)-20 (1).

CLIFF, G (1995): Sharks caught in the protective gill nets off KwaZulu-Natal, South Africa. 8. The great hammerhead shark *Sphyrna mokarran* (Rüppell). *South African Journal of Marine Science*, 15: 105-114.

CLIFF, G and SFJ DUDLEY (1992): Sharks caught in the protective gill nets off Natal, South Africa. 4. The bull shark *Carcharhinus leucas* (Valenciennes). *South African Journal of Marine Science*, 10: 253-270.

CLIFF, G and SFJ DUDLEY (1992): Sharks caught in the protective gill nets off Natal, South Africa. 6. The copper shark *Carcharhinus brachyurus* (Günther). *South African Journal of Marine Science*, 12: 663-674.

CLIFF, G and SFJ DUDLEY (1991): Sharks caught in the protective gill nets off Natal, South Africa. 5. The Java shark *Carcharhinus amboinensis* (Müller and Henle). *South African Journal of Marine Science*, 11: 443-453.

CLIFF, G, DUDLEY, SFJ and B DAVIS (1989): Sharks caught in the protective gill nets off Natal, South Africa. 2. The great white shark *Carcharodon carcharias* (Linnaeus). *South African Journal of Marine Science*, 8: 131-144.

CLIFF, G, DUDLEY, SFJ and B DAVIS (1989): Sharks caught in the protective gill nets off Natal, South Africa. 3. The shortfin mako shark *Isurus oxyrinchus* (Linnaeus). *South African Journal of Marine Science*, 9: 115-126.

CLIFF, G, DUDLEY, SFJ and MR JURY (1996): Catches of white sharks in KwaZulu-Natal, South Africa and environmental influences. Pp. 351-362 in Klimley, AP and DG Ainley (eds): *Great white sharks: The biology of* Carcharodon carcharias. Academic Press, San Diego, 518 pp.

COCKROFT, VG, CLIFF, G and JB ROSS (1989): Shark predation on Indian Ocean bottlenose dolphins *Tursiops truncatus* off Natal, South Africa. *South African Journal of Zoology*, 24 (4): 305-310.

COLLIER, R (2003): *Shark Attacks of the Twentieth Century from the Pacific Coast of North America*. Scientia Publishing, LLC, Chatsworth, 296 pp.

COLLIER, RS, MARKS, M and RW WARNER (1996): White shark attacks on inanimate objects along the Pacific coast of North America. Pp. 217-221 in Klimley, AP and DG Ainley (eds): *Great white sharks: The biology of* Carcharodon carcharias. Academic Press, San Diego, 518 pp.

COMPAGNO, LJV (1984): FAO Species Catalogue. Vol. 4. Sharks of the World. An annotated and illustrated catalogue of shark species known to date. *FAO Fisheries Synopsis*, 125: 1-655.

CONDORELLI, M and GG PERRANDO (1909): Notizie sul *Carcharodon carcharias* L, catturato nelle acque di Augusta e considerazioni medico-legali sui resti umani trovati nel suo tubo digerente. *Bollettino della Società Zoologica Italiana*, 1909: 164-183.

CONNOR, RC and MR HEITHAUS (1996): Approach by great white shark elicits flight response in bottlenose dolphins. *Marine Mammal Science*, 12 (4): 602-606.

CORKERON, PJ, MORRIS, RJ and MM BRYDEN (1987): Interactions between bottlenose dolphins and sharks in Moreton Bay, Queensland. *Aquatic Mammals*, 13: 109-113.

COSTA, F (1991): *Atlante dei Pesci dei mari italiani*. Mursia editrice, Milano, 429 pp.

COSTANTINI, M, BERNARDINI, M, CORDONE, P, GIULIANINI, PG and G OREL (2000): Osservazioni sulla pesca, la biologia riproduttiva ed alimentare di *Mustelus mustelus* (Chondrichtyes, Triakidae) in Alto Adriatico. *Biologia Marina Mediterranea*, 7 (1): 427-432.

COUSTEAU, JP and P COUSTEAU (1970): *The shark: splendid savage of the sea*. Cassell, London.

CROSS, JN (1988): Aspects of the biology of two scyliorhinid sharks, *Apristurus brunneus* and *Parmaturus xaniurus*, from the upper continental slope off southern California. *Fishery Bulletin*, 86 (4): 691-702.

DE MADDALENA, A (2000): Historical and contemporary presence of the great white shark, *Carcharodon carcharias* (Linnaeus, 1758), in the Northern and Central Adriatic Sea. *Annales, Series historia naturalis*, 10 (1): 3-18.

DE MADDALENA, A (2002): *Lo squalo bianco nei mari d'Italia*. Ireco, Formello, 144 pp.

DE MADDALENA, A (2004): Sharks: dangerous or endangered? *The World and I*, 19 (1): 148-155.

DE MADDALENA, A and H BAENSCH (2005): *Haie im Mittelmeer*. Franckh-Kosmos Verlags-GmbH and Co., Stuttgart, 240 pp.

DE MADDALENA, A, PRETI, A and T POLANSKY (2007): *A Guide to the Sharks of the Pacific Northwest (Including Oregon, Washington, British Columbia and Alaska)*. Harbour Publishing, Madeira Park, 160 pp.

DE MADDALENA, A, PRETI, A and R SMITH (2005): *Mako sharks*. Krieger Publishing, Malabar, 72 pp.

DE MADDALENA, A, ZUFFA, M, LIPEJ, L and A CELONA (2001): An analysis of the photographic evidences of the largest great white sharks, *Carcharodon carcharias* (Linnaeus, 1758), captured in the Mediterranean Sea with considerations about the maximum size of the species. *Annales, Series historia naturalis*, 11 (2): 193-206.

DEMSKI, LS and RG NORTHCUTT (1996): The brain and cranial nerves of the white shark: an evolutionary perspective. Pp. 121-130 in Klimley, AP and DG Ainley (eds): *Great white sharks. The biology of* Carcharodon carcharias. Academic Press, San Diego, 518 pp.

DINGERKUS, G (1987): Shark attack in the United States. Pp. 122-133 in Stevens, JD (ed): *Sharks*. Intercontinental Publishing Corporation Limited, Hong Kong, 240 pp.

DINGERKUS, G (1987): Shark distribution. Pp. 36-47 in Stevens, JD (ed): *Sharks*. Intercontinental Publishing Corporation Limited, Hong Kong, 240 pp.

EBERT, DA (1991): Diet of the sevengill shark *Notorynchus cepedianus* in the temperate coastal waters of southern Africa. *South African Journal of Marine Science*, 11: 565-572.

EBERT, DA (1994): Diet of the sixgill shark *Hexanchus griseus* off southern Africa. *South African Journal of Marine Science*, 14: 213-218.

EBERT, DA, COMPAGNO, LJV and PD COWLEY (1992): A preliminary investigation of the feeding ecology of squaloid sharks off the west coast of southern Africa. *South African Journal of Marine Science*, 12: 601-609.

EBERT, DA, COWLEY, PD and LJV COMPAGNO (1996): A preliminary investigation of the feeding ecology of catsharks (Scyliorhinidae) off the west coast of southern Africa. *South African Journal of Marine Science*, 17: 233-240.

ELLIS, R (1983): *The book of sharks*. Robert Hale, London, 256 pp.

ELLIS, R and JE McCOSKER (1991): *Great white shark*. Stanford University Press, Stanford, 270 pp.

GABRIOTTI, V and A DE MADDALENA (2004): Observations of an approach behaviour to a possible prey performed by some great white sharks, *Carcharodon carcharias* (Linnaeus, 1758), at the Neptune Islands, South Australia. *Bollettino del Museo civico di Storia Naturale di Venezia*, 55: 151-157.

GABRIOTTI, V and R GABRIOTTI (2002): Danzando tra le onde. *Sub*, 206: 96-101.

GILBERT, PW (1968): The shark: barbarian and benefactor. *BioScience*, 18 (10): 946-950.

GOLDMAN, KJ, ANDERSON, SD, McCOSKER, JE and AP KLIMLEY (1996): Temperature, swimming depth, and movements of a white shark at the South Farallon Islands, California. Pp. 111-120 in Klimley, AP and DG Ainley (eds): *Great white sharks: The biology of* Carcharodon carcharias. Academic Press, San Diego, 518 pp.

GOTSHALL, DW and T JOW (1965): Sleeper sharks (*Somniosus pacificus*) off Trinidad, California, with life history notes. *California Fish and Game*, 51 (4): 294-298.

GOTTFRIED, MD, COMPAGNO, LJV and SC BOWMAN (1996): Size and skeletal anatomy of the giant megatooth shark *Carcharodon megalodon*. Pp. 55-66 in Klimley, AP and DG Ainley (eds): *Great white sharks. The biology of* Carcharodon carcharias. Academic Press, San Diego, 518 pp.

HODGSON, ES (1987): The shark's senses. Pp. 76-83 in Stevens, JD (ed): *Sharks*. Intercontinental Publishing Corporation Limited, Hong Kong, 240 pp.

HUGHES, R (1987): Shark attack in Australian waters. Pp. 108-121 in Stevens, JD (ed). *Sharks*. Intercontinental Publishing Corporation Limited, Hong Kong, 240 pp.

IRVINE, B, WELLS, RS and PW GILBERT (1973): Conditioning an Atlantic bottlenosed dolphin, *Tursiops truncatus*, to repel various species of sharks. *Journal of Mammology*, 54: 503-505.

JOHNSON, RH (1978): *Sharks of Polynesia*. Les Editions du Pacifique, Papeete, 170 pp.

JOYCE, WN, CAMPANA, SE, NATANSON, LJ, KOHLER, NE, PRATT Jr, HL and CF JENSEN (2002): Analysis of stomach contents of the porbeagle shark (*Lamna nasus* Bonnaterre) in the northwest Atlantic. *ICES Journal of Marine Science*, 59: 1263-1269.

KAJIURA, SM (2001): Head morphology and electrosensory pore distribution of carcharhinid and sphyrnid sharks. *Environmental Biology of Fishes*, 61: 125-133.

KALMIJN, A (1971): The electric sense of sharks and rays. *Journal of Experimental Biology*, 55: 371-383.

KLIMLEY, AP (1994): The predatory behaviour of the white shark. *American Science*, 82: 122-134.

KLIMLEY, AP and SD ANDERSON (1996): Residency patterns of white sharks at the South Farallon Islands, California. Pp. 365-373 in Klimley, AP and DG Ainley (eds): *Great white sharks: The biology of* Carcharodon carcharias. Academic Press, San Diego, 518 pp.

KLIMLEY, AP, ANDERSON, SD, PYLE, P and RP HENDERSON (1992): Spatiotemporal patterns of white shark (*Carcharodon carcharias*) predation at the South Farallon Islands, California. *Copeia*, 1992 (3): 680-690.

KLIMLEY, AP, BUTLER, SB, NELSON, DR and AT STULL (1988): Diel movements of scalloped hammerhead sharks (*Sphyrna lewini* Griffith and Smith) to and from a seamount in the Gulf of California. *Journal of Fish Biology*, 33: 751-761.

KLIMLEY, AP, PYLE, P and SD ANDERSON (1996): Tail slap and breach: agonistic displays among white sharks? Pp. 241-255 in Klimley, AP and DG Ainley (eds): *Great white sharks: The biology of* Carcharodon carcharias. Academic Press, San Diego, 518 pp.

KLIMLEY, AP, PYLE, P and SD ANDERSON (1996): The behaviour of white sharks and their pinniped prey during predatory attacks. Pp. 175-191 in Klimley, AP and DG Ainley (eds): *Great white sharks: The biology of* Carcharodon carcharias. Academic Press, San Diego, 518 pp.

KLIMLEY, AP, LE BOEUF, BJ, CANTARA, KM, RICHERT, JE, DAVIS, SF, VAN SOMMERAN, S and JT KELLY (2001): The hunting strategy of white sharks (*Carcharodon carcharias*) near a seal colony. *Marine Biology*, 138: 617-636.

KUBOTA, T, SHIOBARA, Y and T KUBODERA (1991): Food habits of the frilled shark *Chlamydoselachus anguineus* collected from Suruga Bay, Central Japan. *Nippon Suisan Gakkaishi*, 57 (1): 15-20.

LAST, PR and JD STEVENS (1994): *Sharks and rays of Australia*. CSIRO, Australia, 514 pp.

LE BOEUF, BJ, RIEDMAN, RM and RS KEYES (1982): White shark predation on pinnipeds in California coastal waters. *Fishery Bulletin*, 80: 891-895.

LE BOEUF, BJ and DE CROCKER (1996): Diving behaviour of elephant seals: implications for predator avoidance. Pp. 193-206 in Klimley, AP and DG Ainley (eds): *Great white sharks. The biology of* Carcharodon carcharias. Academic Press, San Diego, 518 pp.

LEVINE, M (1996): Unprovoked attacks by white sharks off the South African coast. Pp. 435-448 in Klimley, AP and DG Ainley (eds): *Great white sharks. The biology of* Carcharodon carcharias. Academic Press, San Diego, 518 pp.

LINEAWEAVER III, TH and RH BACKUS (1969): *The Natural History of sharks*. JB Lippincott Co., Philadelphia, 256 pp.

LINNAEUS, C (1758): *Systema naturae. 10th Edition. Vol. 1, Regnum animale*. Salvii, Stockholm, 824 pp.

LIPEJ, L, DE MADDALENA, A and A SOLDO (2004): *Sharks of the Adriatic Sea*. Knjiznica Annales Majora, Koper, 254 pp.

LONG, DJ, HANNI, KD, PYLE, P, ROLETTO, J, JONES, RE and R BANDAR (1996): White shark predation on four pinniped species in central California waters: geographic and temporal patterns inferred from wounded carcasses. Pp. 263-274 in Klimley, AP and DG Ainley (eds): *Great white sharks. The biology of* Carcharodon carcharias. Academic Press, San Diego, 518 pp.

LONG, DJ and RE JONES (1996): White shark predation and scavenging on cetaceans in the eastern north Pacific Ocean. Pp. 293-307 in Klimley, AP and DG Ainley (eds): *Great white sharks. The biology of* Carcharodon carcharias. Academic Press, San Diego, 518 pp.

LONG, JA (1995): *The rise of fishes: 500 million years of Evolution.* John Hopkins University Press, Baltimore, 224 pp.

LOPEZ A, BARRULL, J and I MATE (1996): Feridas nos corpos dos cetaceos varados en Galicia como indicio de depredación e de alimentación baseada en preas, principalmente por quenllas. *Eubalaena*, 9: 10-21.

LOWE, CG (2001): Metabolic rates of juvenile scalloped hammerhead sharks (*Sphyrna lewini*). *Marine Biology*, 139: 447-453.

LYLE, JM (1983): Food and feeding habits of the lesser spotted dogfish, *Scyliorhinus canicula* (L.), in Isle of Man waters. *Journal of Fish Biology*, 23 (6): 725-737.

MAISEY, JG (1987): Evolution of the shark. Pp. 14-16 in Stevens, JD (ed): *Sharks.* Intercontinental Publishing Corporation Limited, Hong Kong, 240 pp.

MARSHALL, TC (1965): *Fishes of the Great Barrier Reef and coastal waters of Queensland.* Livingston Publishing Company, 566 pp.

MARTIN, RA (1992): Cold fire in the sea. *DIVER Magazine*, June 1992: 18-19.

MARTIN, RA (1995): *Shark smart: the divers' guide to understanding shark behaviour.* Diving Naturalist Press, Vancouver, 180 pp.

MARTIN, RA (2003): *Field Guide to the Great White Shark.* ReefQuest Centre for Shark Research, Special Publication No. 1, 192 pp.

MASUDA, R and DA ZIEMANN (2003): Vulnerability of Pacific threadfin juveniles to predation by bluefin trevally and hammerhead shark: size dependent mortality and handling stress. *Aquaculture*, 217: 249 257.

MATTHIESSEN, P (1971): *Blue meridian* Random House, New York, 204 pp.

MELVILLE, H (1851): *Moby Dick.* Random House, New York, 822 pp.

McCOSKER, JE (1987): The white shark, *Carcharodon carcharias*, has a warm stomach. *Copeia*, 1987: 195-197.

MICHAEL, SW (1993): *Reef sharks and rays of the world.* Sea Challengers, Monterey, 107 pp.

MOLLET, HF, CAILLIET, GM, KLIMLEY, AP, EBERT, DA, TESTI, AD and LJV COMPAGNO (1996): A review of length validation methods and protocols to measure large white sharks. Pp. 91-108 in Klimley, AP and DG Ainley (eds): *Great white sharks. The biology of* Carcharodon carcharias. Academic Press, San Diego, 518 pp.

MORENO, JA (1995): *Guía de los tiburones de aguas ibéricas, Atlántico Nororiental y Mediterráneo.* Ediciones Pirámide, Madrid, 310 pp.

MOTTA, PJ, HUETER, RE, TRICAS, TC and AP SUMMERS (2003): Kinematic analysis of suction feeding in the nurse shark, *Ginglymostoma cirratum* (Orectolobiformes, Ginglymostomatidae). *Copeia*, 2002 (1): 24-38.

MUNTHE, A (1928): *The story of San Michele.* Butler and Tanner Ltd, London.

MYRBERG Jr, A (1987): Shark behaviour. Pp. 84-92 in Stevens, JD (ed): *Sharks.* Intercontinental Publishing Corporation Limited, Hong Kong, 240 pp.

NAGASAWA, K (1998): Predation by salmon sharks (*Lamna ditropis*) on Pacific salmon (*Oncorhynchus* spp.) in the North Pacific Ocean. *North Pacific Anadromous Fish Commission*, 2: 419-433.

NAKAYA, K (1991): A review of the long-snouted species of *Apristurus* (Chondrichthyes, Scyliorhinidae). *Copeia*, 1991 (4): 992-1002.

OLSEN, AM (1987): Using sharks. Pp. 186-193 in Stevens, JD (ed): *Sharks.* Intercontinental Publishing Corporation Limited, Hong Kong, 240 pp.

PAUL, LJ (1987): Shark attack in New Zealand. Pp. 148-155 in Stevens, JD (ed): *Sharks.* Intercontinental Publishing Corporation Limited, Hong Kong, 240 pp.

PISCITELLI, L and A DE MADDALENA (2005): Evidence of a predatory attack on a large paromola, *Paromola cuvieri* (Risso, 1816), by a kitefin shark, *Dalatias licha* (Bonnaterre, 1788). *Thalassia Salentina*, 28: 3-8.

PYLE, P and S ANDERSON (2002): *White shark research at Southeast Farallon Island, 2001. Report to the US Fish and Wildlife Service.* Farallon National Wildlife Refuge, 7 pp.

PYLE, P, ANDERSON, SD, KLIMLEY, AP and RP HENDERSON (1996): Environmental factors affecting the occurrence and behaviour of white sharks at the Farallon Islands, California. Pp. 281-291 in Klimley, AP and DG Ainley (eds): *Great white sharks. The biology of* Carcharodon carcharias. Academic Press, San Diego, 518 pp.

POWELL, DC (2001): A fascination for fish. Adventures of an underwater pioneer. UC Press / Monterey Bay Aquarium Series in Marine Conservation, 3, 339 pp.

PRATT, HL, CASEY, JG and RB CONKLIN (1982): Observations on large white sharks, *Carcharodon carcharias*, off Long Island, New York. *Fishery Bulletin*, 80: 153-156.

PRETI, A, SMITH, SE and DA RAMON (2001): Feeding habits of the common thresher shark (*Alopias vulpinus*) sampled from the California-based drift gill net fishery, 1998-99. *California Cooperative Oceanic Fisheries Investigations Reports*, 42: 145-152.

RANDALL, JE (1986): *Sharks of Arabia.* IMMEL Publishing, London, 148 pp.

READER'S DIGEST (1992): *Sharks. Silent hunters of the deep.* Reader's Digest Australia, Surry Hills, 208 pp.

RONDELET, G (1554): *Libri de Piscibus Marinis, in quibus verae Piscium effigies expressae sunt.* Bonhomme, Lyon, 583 pp.

SCIARROTTA, TC and DR NELSON (1977): Diel behaviour of the blue shark, *Prionace glauca*, near Santa Catalina Island, California. *Fishery Bulletin*, 75 (3): 519-528.

SCOTT, TD (1962): *The marine and fresh water fishes of South Australia.* WL Hawes, Government Printer, Adelaide.

SHAW, C (2000): *Sacred Monkey River. A canoe trip with the Gods.* Norton, New York, 316 pp.

SIMS, DW and VA QUAYLE (1998): Selective foraging behaviour of basking sharks on zooplankton in a small-scale front. *Nature*, 393: 460- 464.

SMALE, MJ (1991): Occurrence and feeding of the three shark species, *Carcharhinus brachyurus, C. obscurus* and *Sphyrna zygaena*, on the eastern Cape coast of South Africa. *South African Journal of Marine Science*, 11: 31-42.

SMITH, JLB (1949): *The sea fishes of Southern Africa.* Central News Agency, Johannesburg, 550 pp.

SNYDERMAN, M (1987): Do sea lions eat sharks? P. 95 in Stevens, JD (ed): *Sharks.* Intercontinental Publishing Corporation Limited, Hong Kong, 240 pp.

SNYDERMAN, M (1987): When sharks and dolphins cross paths. P. 103 in Stevens, JD (ed): *Sharks.* Intercontinental Publishing Corporation Limited, Hong Kong, 240 pp.

SPRINGER, S (1960): Natural history of the sandbar shark (*Eulamia milberti*). *Fishery Bulletin*, 61 (178): 1-38.

STEEL, R (1985): *Sharks of the world.* Blanford Press, Poole (Dorset), 192 pp.

STEVENS, JD (1984): Biological observations on sharks caught by sport fishermen off New South Wales. *Australian Journal Marine Freshwater Resources*, 35: 573-590.

STEVENS, JD (1987): Shark biology. Pp. 50-75 in Stevens, JD (ed): *Sharks.* Intercontinental Publishing Corporation Limited, Hong Kong, 240 pp.

STEVENS, JD (ed) (1987): *Sharks.* Intercontinental Publishing Corporation Limited, Hong Kong, 240 pp.

STEWART, BS and PK YOCHEM (1984): Radio-tagged harbor seal, *Phoca vitulina richardsi*, eaten by a white shark, *Carcharodon carcharias*, in the Southern California Bight. *California Fish and Game*, 71: 113-115.

STILLWELL, C (1991): The ravenous mako. Pp. 77-88 in Gruber, SH (ed): Discovering sharks. *Underwater Naturalist, Bulletin American Littoral Society*, 19 (4)-20 (1).

STORAI, T, ZUFFA, M and R GIOIA (2001): Evidenze di predazione su odontoceti da parte di *Isurus oxyrinchus* (Rafinesque, 1810) nel Tirreno Meridionale e Mar Ionio (Mediterraneo). *Atti Società toscana Scienze naturali*, Mem., Serie B, 108: 71-75.

STORER, TI and RL USINGER (1965): *General Zoology*, 4th Edition. McGraw-Hill, New York, 741 pp.

STRONG Jr, WR (1991): Instruments of natural selection: how important are sharks? Pp. 70-73 in Gruber, SH (ed): Discovering sharks. *Underwater Naturalist, Bulletin American Littoral Society*, 19 (4)-20 (1).

STRONG Jr, WR (1996): Repetitive aerial gaping: a thwart-induced behaviour in white sharks. Pp. 207-215 in Klimley, AP and DG Ainley (eds): *Great white sharks. The biology of* Carcharodon carcharias. Academic Press, San Diego, 518 pp.

STRONG Jr, WR (1996): Shape discrimination and visual predatory tactics in white sharks. Pp. 229-240 in Klimley, AP and DG Ainley (eds): *Great white sharks. The biology of* Carcharodon carcharias. Academic Press, San Diego, 518 pp.

STRONG, WR, BRUCE, BD, MURPHY, RC and DR NELSON (1992): Movements and associated observations of bait-attracted white sharks, *Carcharodon carcharias*: a preliminary report. *Australian Journal Marine Freshwater Resources*, 43: 13-40.

TAYLOR, CK and GS SAAYMAN (1973): The social organization and behaviour of bottlenose dolphins (*Tursiops aduncus*) and baboons (*Papio ursinus*): some comparisons and assessments. *Annals of the Cape Provincial Museums (Natural History)*, 9 (2): 11-49.

TRICAS, TC (1987): Shark ecology. Pp. 96-101 in Stevens, JD (ed): *Sharks*. Intercontinental Publishing Corporation Limited, Hong Kong, 240 pp.

TRICAS, TC and JE McCOSKER (1984): Predatory behaviour of the white shark (*Carcharodon carcharias*) with notes on its biology. *Proceedings of the California Academy of Sciences*, 43 (14): 221-238.

URQUHART, DL (1981): The North Pacific salmon shark. *Sea Frontiers*, November-December: 361-363.

WATTS, S (2001): *The end of the line?* WildAid, San Francisco, 62 pp.

WELLS, RS (1991): Bringing up baby. *Natural History*, August: 56-62.

WEST, J (1996): White shark attacks in Australian waters. Pp. 449-455 in Klimley, AP and DG Ainley (eds): *Great white sharks: The biology of* Carcharodon carcharias. Academic Press, San Diego, 518 pp.

WETHERBEE, B (1991): Feeding biology of sharks. Pp. 74-76 in Gruber, SH (ed): Discovering sharks. *Underwater Naturalist, Bulletin American Littoral Society*, 19 (4)-20 (1).

YANG, M-S and BN PAGE (1999): Diet of Pacific sleeper shark, *Somniosus pacificus*, in the Gulf of Alaska. *Fishery Bulletin*, 97: 406-409.

YANO, K, YABUMOTO, Y, TANAKA, S, TSUKADA, O and M FURUTA (1999): Capture of a mature female megamouth shark, *Megachasma pelagios*, from Mie, Japan. Pp. 335-349 in Séret, B and J-Y Sire (eds): Proceedings of the 5th Indo-Pacific Fish Conference, Nouméa, 1997. Societé Française d'Ichtyologie, Paris.

Shark species index

A

Alopias superciliosus (see bigeye thresher)
Alopias vulpinus (see common thresher shark)
Alopiidae 113, 141, 150, 183
Apristurus brunneus (see brown catshark)
Apristurus herklotsi (see longfin catshark)
Apristurus longicephalus (see longhead catshark)
Apristurus microps (see smalleye catshark)

B

barbeled catshark 131, 183
basking shark 13, 22, 34, 141, 146–148, 152, 164, 183
bigeye thresher 13, 79, 142, 183
black dogfish 70, 181
blacknose shark 88, 91, 167, 184
blackspotted smooth-hound 139, 184
blacktip reef shark 18, 52, 73, 86, 109, 142, 157, 163, 184
blacktip shark 21, 22, 71, 86, 88, 143, 150, 151, 154, 184
blind shark 92, 182
blue shark 15, 22, 36, 39, 44, 45, 48, 49, 53, 70, 71, 78, 80, 85, 89, 90, 92, 108, 141–143, 153–156, 163, 167, 171, 175, 184, 192
bluntnose sixgill shark 13, 39, 53, 54, 70, 71, 77, 79, 101, 109, 126, 154, 181, 186
bonnethead 53, 66, 70, 91, 112, 167, 184
Brachaelurus waddi (see blind shark)
broadnose sevengill shark 71, 79, 92, 100, 109, 116, 154, 181
bronze whaler shark 68, 70, 71, 76, 80, 86, 92, 144, 154, 184
brown catshark 70, 183
brown smooth-hound 136, 184
bull shark 10, 22, 44, 65, 69, 70–73, 78, 84, 89, 92, 93, 95, 100, 101, 145, 154, 184, 187

C

Carcharhinidae 53, 65, 78, 88, 109, 150, 153, 168, 174, 184
Carcharhiniformes 11
Carcharhinus acronotus (see blacknose shark)
Carcharhinus albimarginatus (see silvertip shark)
Carcharhinus amblyrhynchos (see grey reef shark)
Carcharhinus amboinensis (see pigeye shark)
Carcharhinus brachyurus (see bronze whaler shark)
Carcharhinus brevipinna (see spinner shark)
Carcharhinus falciformis (see silky shark) *Carcharhinus galapagensis* (see Galapagos shark)
Carcharhinus leucas (see bull shark)
Carcharhinus limbatus (see blacktip shark)
Carcharhinus longimanus (see oceanic whitetip shark)
Carcharhinus melanopterus (see blacktip reef shark)
Carcharhinus obscurus (see dusky shark)
Carcharhinus perezi (see Caribbean reef shark)
Carcharhinus plumbeus (see sandbar shark)
Carcharias taurus (see sandtiger shark)
Carcharodon carcharias (see great white shark)
Carcharodon megalodon (see megatooth shark)
Caribbean reef shark 86, 89, 184
Centrophorus granulosus (see gulper shark)
Centrophorus squamosus (see leafscale gulper shark)
Centroscyllium fabricii (see black dogfish)
Centroscymnus coelolepis (see Portuguese shark)
Cephaloscyllium isabellum (see draughtsboard shark)
Cephaloscyllium ventriosum (see swellshark)
Cetorhinidae 183
Cetorhinus maximus (see basking shark)

195